IMAGING SATURN

OTHER BOOKS BY HENRY S. F. COOPER, JR.

Apollo on the Moon

Moon Rocks

13: The Flight That Failed

A House in Space

The Search for Life on Mars

IMAGING

THE VOYAGER FLIGHTS TO SATURN

SATURN

HENRY S. F. COOPER, JR.

HOLT, RINEHART AND WINSTON
NEW YORK

Published by Holt, Rinehart and Winston,
383 Madison Avenue, New York, New York 10017.
Published simultaneously in Canada by Holt, Rinehart
and Winston of Canada, Limited.

Library of Congress Cataloging in Publication Data
Cooper, Henry S. F.
Imaging Saturn.
1. Project Voyager. 2. Saturn probes. 3. Saturn
(Planet)—Photographs from space. 4. Imaging systems
in astronomy. I. Title.
QB671.C66 1982 523.4'6 83-10
ISBN: 0-03-061688-3

First Edition
Designer: Elissa Ichiyasu
Printed in the United States of America
10 9 8 7 6 5 4 3 2 1

Grateful acknowledgment is made to NASA
for all of the photographs used in this work.
Most of this book originally appeared in slightly
different form in *The New Yorker*.

ISBN 0-03-061688-3

In memory of William L. Cary

1910–1983

CONTENTS

A section of color photographs follows page 50.

PREFACE

Of the space program, it has been said that it has expanded mankind's environment—that is, his home turf, as defined by places he himself has been or has physically sent his machines. For a long, long time, man's environment was limited to his home planet, the Earth; then he ventured to the Moon, a quarter of a million miles away; then he sent his spacecraft to the other inner planets—actually landing on Mars, which orbits thirty-five million miles and more beyond the Earth. After that, we extended our presence out to Jupiter, some four hundred million miles beyond the Earth, and now to Saturn, half a billion miles beyond that. Our place in the sun has dramatically and suddenly extended outward like ripples in a pond, encompassing some of the farthest orbits in the solar system.

A person does not necessarily connect with the manner in which mankind's physical domain has grown, unless he becomes involved with the new places—which he might do, among other ways, by writing about them. It wasn't until I was correcting the final manuscript of this book about the Voyager missions to Saturn, its rings, and its moons, and came across a reference I had made to the Martian moon Phobos that I realized how far we had come: when in an earlier book, about the Viking landings on Mars, I had written about Phobos, it had seemed at the end, certainly, of *my* universe, and probably a lot of other people's as well; now it seemed as familiar a landmark as the worn, exposed boulder in the field outside my study window. It is satisfying to have your sense of place extended so far without having to go to those cold, inhospitable spots yourself. Perhaps Saturn will seem as familiar a landmark after Voyager 2 has traveled on to Uranus, and then to Neptune—which will remain, as far as mankind is concerned, the Ultima Thule of his sphere for some time to come. In the meantime, I hope reading this book will help you lay a claim on some of the intervening territories—as writing it has done for me.

In the course of writing it, many people have helped me. First, of course, are the members of the Voyager imaging team—the group that worked with the pictures—and in particular its leader, Bradford A. Smith, who initially made me welcome. I have naturally concentrated on a few members of the team; other books could have been written about any other combination of them, for each scientist contributed to the analysis of the Voyager pictures in his, or her, own unique way. All have lent me a hand and have put up with my prowling about at inconvenient times, asking all sorts of nuisancy questions. Many team members have continued to be helpful, for longer perhaps than they had anticipated, answering further questions and straightening out parts of the manuscript—in particular Bradford Smith, but also Reta Beebe, Jeffrey Cuzzi, Andrew Ingersoll, Harold Masursky, Tobias Owen, Eugene Shoemaker, and Laurence Soderblom. Richard Terrile deserves special thanks as my unofficial docent. And so does David Morrison, for sharing with me the manuscript of his own excellent account of the mission, handsomely published by the National Aeronautics and Space Administration under the title *Voyages to Saturn* (NASA SP-451).

Other, nonscientific, people have helped with this book, too, in such ways as opening doors for me; and, later, by blocking metaphors, correcting syntax, and checking facts. At the Jet Propulsion Laboratory in Pasadena, I want to thank in particular Frank Bristow, who not only opened doors but threw wide the main gate; Don Bane and Mary Beth Murill, who supplied me with information; and Jurrie van der Woude, who tracked down all the pictures used in this book. I also want to thank the staff of *The New Yorker*, in whose pages much of the book first appeared—in particular, Sara Lippincott, John Bennet, and Nancy Franklin—as well as my editors at Holt, Rinehart and Winston, Donald Hutter and Marian Wood.

Henry S. F. Cooper, Jr.
COOPERSTOWN, N.Y.
SEPTEMBER, 1982

PART I

VOYAGER 1

NOVEMBER 1980

I N B O U N D

Saturn, the most distant of the planets visible to the ancients—it is some nine hundred million miles from the Sun—has long been considered the most beautiful and mysterious of them all. This has been so possibly ever since 1610, when Galileo Galilei trained a thirty-two-power telescope upon it and noticed what he later called an *ansa* (or "handle") protruding from each side, and certainly since 1656, when the Dutch astronomer Christiaan Huygens, using a hundred-power telescope, identified those handles as a separate, encircling ring. Ever since, Saturn and its rings and moons have been wondrous to behold—around a soft, shimmering, celestial O, a sharp parenthesis of hard, white light, itself surrounded by luminous, tiny dots, thus:

···(O)···

Galileo and Huygens had no idea what the rings were composed of or how they had formed—or, perhaps more important, what held them in place. Probably no other celestial body within view demonstrates so dramatically the fact that the behavior of objects in space is different from what we are used to in the atmosphere and gravity of the Earth: rings could not be orbited around any terrestrial object—a tree, a mountain—because the Earth's atmosphere would slow them down and the Earth's gravity, greater than the tree's or the mountain's, would pull them down. Far more than any other known planet or star, therefore, Saturn and its rings—until very recently the only known ones and still the biggest and brightest—have been a symbol of space.

In November, 1980, when the spacecraft Voyager 1 was scheduled to rendezvous with Saturn—Voyager 1 was the first half of a two-part mission, with a duplicate spacecraft, Voyager 2, due to encounter Saturn nine months later, on August 25, 1981—I went out to Pasadena, California, to the Jet Propulsion Laboratory, which was running the mission for the National Aeronautics and Space Administration. There I spent some time with scientists as they analyzed the data arriving from the planet. A flyby mission, which neither lands nor orbits but continues on its way, may provide the best opportunity there is to observe the process of scientific discovery, for the time and place are known beforehand; the duration is short; the scientists, like runners on their mark, are poised to make new observations; and something new is almost guaranteed to turn up. Just as explorers had kept journals in the past, I wanted to chronicle this expedition, following the way old ideas gave way to new ones as the data came in, and recording even the wrong turns, which are as much a part of discovery as the right ones. Many of the scientists involved regarded the Voyager missions as the culmination of the great voyages of discovery beyond the Earth that began, for this country, with the first successful Ranger missions to the Moon in 1964, and for the Soviet Union with the Luna series, whose first successes were in 1959. Before that time, mankind knew well only one body in the solar system—the Earth. By 1976, thanks in part to the Viking landings on Mars that year and the Soviet Union's Venus missions of 1975, the number had been increased to seven, consisting of the four inner, rocky planets and their moons. Then, when Voyager 1 reached Jupiter and its sixteen moons, in March, 1979, that number was more than doubled, and when it reached Saturn, with twelve known moons and, doubtless, even more unknown moonlets, it was expected to double again (and indeed did). Voyager 2, which unlike Voyager 1 will go on to encounter Uranus four and a half years after leaving Saturn, and Neptune three years and eight months after that, will bring the total number of planets and moons now known to at least fifty-five, all but the first seven inspected by the Voyager mission. (Pluto, a small rocky or icy body which orbits beyond Neptune, will not be visited by Voyager 2.) Very few of these new worlds are of the rocky variety. The four giant outer planets are largely gaseous and their forty-odd moons are mixtures of ice and rock, with the ice predominating in most cases. Not only are the outer planets very different from the inner ones but the moons of Jupiter had exhibited a number of surprising characteristics, and the scientists believed that Saturn's would as well.

Because Voyager carried many scientific instruments and experiments, the scientists—about a hundred and twenty-five in all and collectively called the science team—had been divided into separate teams, such as the infrared-radiation team, the ultraviolet-spectroscopy team, the radio-science team, the planetary–radio-astronomy team, the plasma-waves team, and the imaging-science team. There were eleven teams in all under the direction of the chief scientist, Edward C. Stone. I was particularly interested in the data that was being studied by the imaging team. Imaging is a process whereby impulses from television cameras aboard a spacecraft are encoded by computer into a series of numbers and transmitted back to Earth, where they are decoded by another computer and reconstituted into pictures. The scientists had fewer preconceptions about Saturn than they had had about any of the new worlds visited by spacecraft so far, and therefore this would be planetary discovery at its purest. Mars and our Moon had generated hundreds of years of speculation, but with Saturn the scientists were coming in almost cold—and hence there was little of the type of argument among them that had characterized the days before spacecraft had approached our Moon or Mars. Although another spacecraft, Pioneer 11, had flown by Saturn in 1979, its cameras had been vastly inferior; they had made out a few details of the ring system but revealed almost nothing about the moons. One imaging-team member said to me, "Imagine the kick of visiting a whole new planetary system for what amounts to the first time." I was eager to see that system unfold before the scientists' eyes.

On the morning of Sunday, November 9th, when I arrived at the Jet Propulsion Laboratory, or JPL, the space-center seemed deserted. Its grounds cascade level by level down the foothills of the San Gabriel Mountains in a picturesque fall of buildings, some low and sprawling, others narrow mini-skyscrapers, the whole punctuated by pines and ponds. According to a big bulletin board standing in the middle of the central mall, at 8:00 A.M. (P.S.T.) that day the spacecraft was 947,752,700 miles from the Earth and 2,976,707 miles from Saturn and was closing with

Saturn at 34,978 miles per hour, a speed that would increase rapidly under Saturn's gravitational pull; a radio message (computerized data, including data for images) from the spacecraft at that distance would take one hour and twenty-four minutes and fifty seconds to reach Mission Control at JPL; the days to go until the encounter were three.

Saturn offered more objects to look at than did Jupiter, which Voyager 1 had flown by a year and eight months before, and, because the Saturn system was more compact, there would be only a third as much time to look. Before my visit, I had had a glimpse of what was coming: a movie put together by a computer from several score still pictures taken the previous month, beginning when the spacecraft was still some thirty million miles away and moving closer, in what was known as the observatory stage of the mission. In the movie, which everyone called the "zoom" movie as Saturn got progressively bigger, the Saturnian system—the planet with its rings and most of its moons spread out on the plane of its equator—could be seen at a glance. Although Saturn was rotating, its surface was so bland and unmottled that it was hard to fix on any features and so get a sense of its motion. The rotation of one of the major rings—the broadest and brightest, called the B ring—was obvious, however, because it bore dark, fingerlike radial smudges, not seen by Pioneer 11, which the surprised scientists referred to as "spokes," and whose movement around the planet was unmistakable. Several bright specks—the larger moons—could be seen flitting around Saturn like fireflies, the inner ones darting very fast. It was a miniature solar system come alive. (Indeed, it is as a sort of miniature solar system that many scientists regard not only Saturn's system but Jupiter's as well.) In addition to the nine moons known before the Pioneer and Voyager missions—they are, from the outside in, Phoebe, Iapetus, Hyperion, Titan, Rhea, Dione, Tethys, Enceladus, and Mimas—five very small moons, as yet nameless, had been discovered recently (three of them by Voyager), orbiting much closer to the planet. A sixth, sharing the orbit of Dione and temporarily named Dione B, had been identified by ground-based observation a few months before the Voyager discoveries, as had a seventh and an eighth close to Tethys. Saturn's moons are composed largely of water ice and, with the exception of Titan (far and away the biggest, discovered by Huygens in the mid-seventeenth century), are much smaller than our Moon; a few, including Phoebe and the newly discovered ones, are roughly a twentieth its size.

Imaging the moons had already begun by the time I arrived at JPL, even though the closest approaches to them—what NASA calls the "encounters"—would be occurring over the next few days. NASA uses that

Saturn, its rings, and some of its moons, showing as specks outside the rings.

term loosely; since the moons were, of course, at different points on their orbits, at closest approach they would be at greatly varying distances from the spacecraft's trajectory. Phoebe, for example, would never be nearer than eight million miles away, nor show up as more than a point of light. On Tuesday night would be the closest encounter of all, when Voyager would come within twenty-five hundred miles of the haze-shrouded surface of Titan, the only moon in the solar system to possess an atmosphere. Because of its dense clouds, its precise radius was not known, but Titan was thought to be the biggest moon in the solar system—certainly on the same order as the biggest Jovian satellite, Ganymede, and larger than the planet Mercury—though its precise diameter was not known because so far it had been impossible to find Titan's surface beneath its thick covering of cloud. Less than two hours after encountering the giant moon, Voyager would pass beneath Saturn's ring plane. The following day, it would image some of the newly discovered moons; that afternoon, after its closest approach to Tethys, it would swing beneath Saturn and, at 3:45 P.M. (P.S.T.), make its closest approach to the planet, coming within 77,200 miles of its cloud decks. A few hours later, the closest approaches to Mimas, Enceladus, and Dione would occur, and then, at 9:45 P.M., the spacecraft would recross the equatorial plane. After encountering Rhea, Hyperion, and Iapetus, it would proceed up and out of the solar system. The decision to more or less jettison Voyager 1 after it left Saturn had stemmed from the scientists' extreme interest in having it pass close to Titan: because the Saturn system is tilted toward the ecliptic—the plane formed by the paths of all the planets except Pluto as they orbit the Sun—and because at the time of the encounter Titan would be at the low side of the tilt, a route passing close to it *and* around Saturn would necessarily hurl the spacecraft away from the plane the planets are on. The specific justification for this sacrifice was the opportunity it provided to examine Titan's thick brownish-yellow atmosphere, which was known to contain methane, an organic compound, and which therefore might yield information about the development of organic chemistry, the carbon compounds on which life is based, on other bodies—including the Earth. Voyager 2, unlike Voyager 1, was scheduled to remain within the plane of the ecliptic, and so could go on to Uranus, and Neptune, before it, too, left the solar system.

The imaging team's offices were in a bronze-and-glass structure on the mall known as the Science Building. In a corner office on the third floor, I found the imaging team's leader, Bradford A. Smith. An astronomer with the Department of Planetary Sciences at the University of Arizona, in Tucson, Smith is a big man in his late forties, with broad shoulders, a smooth, round face, a firm jaw, and a full mustache. He was studying a number of black-and-white pictures on his desk, deciding which would be printed up in color for release to journalists. The press, he said, was particularly interested in the encounter not only because Saturn was the loveliest of all the planets but because Voyager would be the last planetary mission for some time. The next planned mission—Galileo, to be launched to Jupiter, possibly from the space shuttle in 1985—was not as definite as NASA would like, because of delays in the shuttle program; the status of a Venus orbital mission was uncertain; and a proposal to rendezvous with Halley's Comet in 1986 had not been approved by Congress. Also, it was by no means certain that Voyager 2's instruments, designed to last through Saturn, would continue to work during the proposed Uranus encounter. For seventeen years, beginning in 1964, when the Ranger 7 spacecraft impacted on the Moon, NASA had been getting new information from some moon or planet at least once a year; that sequence was about to end. Hence there was nostalgia—even bitterness—underlying the elation at Voyager's success, for many scientists perceived the Voyager mission as the end of an era.

To turn our thoughts to something happier, I asked Smith when he had first seen Saturn.

"When I was in junior high school, back in Winchester, Massachusetts," he said. "One evening, I rode my bike over to the Harvard College Observatory. By the standards of the observatory of the University of Arizona, where I work now, the seeing was not so great. But there is something about Saturn—even if the seeing is bad—that makes it hard to believe that it's real, that there really is a celestial object that looks like that. No television picture or movie or photograph can capture Saturn. You have to see it through a telescope. You never forget it—ever after, it remains a sort of focal point in the sky."

The imaging team, Smith said, consisted of twenty-six people, including astronomers, geologists, meteorologists, and physicists. They were grouped, somewhat informally, by interest: ring people, planet people, and satellite people, with a smattering of people concentrating on Titan. There was some overlap: the planet people, for instance, tended to be interested in Titan, because Titan, like Saturn itself, involved a good deal of atmospheric physics; and practically everyone was interested in the rings. Smith said that during the encounter period, discoveries made by any member of the imaging team would probably not be claimed individually but would be credited to the entire team. With so many people looking at the pictures more or less at the same time, he said, it was hard to see how matters could be handled differently, and he expected there would be complete freedom in the way of give and take. The excitement was just about to start; Voyager's cameras were now close enough to begin making out details of the ring system and features on some of the moons. Indeed, it had started already.

"Right now, I'm most perplexed about the spokes—the dark radial smudges on the B ring that we first noticed about a month ago," he commented. "They shouldn't be there! Yet there they are, running ten thousand miles across the B ring toward Saturn. They seem to be strongest where the ring emerges from Saturn's shadow, and dissipate as they go around the sunny side. They must be imprinted in the shadow—and imprinted almost instantaneously—by a process we don't understand. We have no theories about the spokes. I've worked on many missions, but I can't think of any other instance where we have been kept puzzled for so long. We're no further along than we were the day they were found."

Smith did not want to hazard any guesses about what the moons might be like. "We were burned so badly on the Jupiter satellites, which turned out to be nothing like what we had expected, that we don't want to stick our necks out again," he said. "Practically everything we'll be seeing from now on in will be new."

Outside Smith's office, a dozen imaging-team members stood gazing up at a television monitor in the main room of the imaging-team's quarters—a gray, square central hall lined with offices. The monitor showed pictures, or "raws," as they were arriving from the spacecraft. The raws were washed-out and murky—they would be enhanced for contrast and detail later—but they were nevertheless dramatic. Saturn was close enough to the spacecraft now to more than fill the screen: a quarter of the planet loomed to the left, in a vast arc; cutting across it, coming apparently from nowhere, was the vast sweep of a section of the rings, curving like some California freeway that was held up by nothing at all. Inside the curves, one could see concentric parallel lines. They resembled tire marks. The whole thing might have been some sensational ride at a fair.

An L-shaped corridor led to a half-dozen more offices, each with desk space for two or three scientists. Most of the office doors were shut, but the doors to two other rooms were open, and the low sounds of intense conversation were coming from within. These rooms were dark and windowless. Each of them contained what is called an interactive—a computer terminal with a television screen on which, simply by pressing buttons on a keyboard, one can call up images and enhance them or enlarge them or enlarge any section of them. Because the raws were of little use to the scientists, they generally preferred to look at the pictures on the interactives. For this, they had to wait until sixty pictures could be gathered together and stored in a computer disc pack. (Though the interactives provided the best early look at the pictures, far better pictures would be available after the images had been through JPL's Image Processing Laboratory, or IPL—something that would take time.)

In the first interactive room, I found Harold Masursky, of the United States Geological Survey's Branch of Astrogeologic Studies, in Flagstaff, Arizona—a man in his mid-fifties with graying hair that, starting from a high forehead, swept back in a striking manner almost to his collar. He was wearing a rumpled white shirt and no tie. While several other team members watched, he punched the computer keyboard to enhance a picture on the screen. The picture, taken early that morning, showed a smooth white disc with what appeared to be a dim spot on it. The disc was the moon Rhea, some nine hundred miles in diameter, seen from

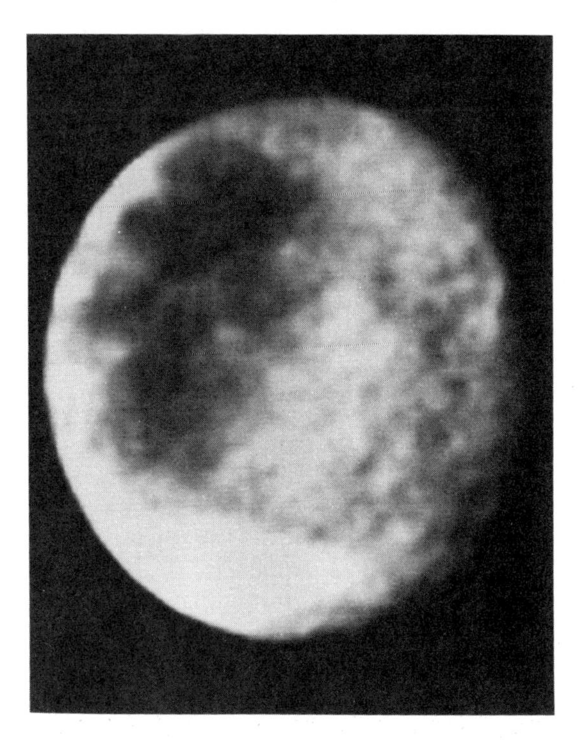

The rosette on Rhea, probably an impact crater; the first feature identified on any of Saturn's moons.

three million miles away—thirteen times the distance of our Moon from the Earth, and too far for anyone to make out much of anything. Masursky was trying to enhance the picture in order to bring out the dim spot. With all the instructions punched in, he pressed a final button, and the enlarged, enhanced picture popped onto the screen. The dim spot had become a bright spot with rays extending from it. "There!" Masursky said. "That's the first identifiable feature seen on any Saturn moon. It's not apt to be a volcano; it's far more likely to be an impact crater. It looks like a rosette." He pulled a lever in order to make a print—a "hard copy," he called it—of the picture. Then he pressed more buttons and began flipping through other pictures in the disc pack.

I asked Masursky, a field geologist, whether astrogeology was much different from geology on Earth. He shook his head. "Whenever you're in the field, wherever you are—on Earth, or looking at photographs of the Moon, or Mars, or the moons of Saturn—there is the same idea of considering an area with an open mind; you start out never knowing what you're going to run into," he said. "It's a nineteenth-century subject, in the sense that—despite our occasional use of computers—it is not computerized or narrow. You have to have a strong intuitive sense of what you are going to see if you go over the next hill; and this is equally true on an Arizona desert or on Rhea. I try to compare features on different planets or moons. Although the moons of Jupiter, which Voyager One encountered a year and a half ago, are composed mainly of ice and have no basalts, like our own Moon, the ice at these cold temperatures behaves a little like lunar basalts, when it comes to fracturing and the effects of impacts. I suspect the same will be true of Saturn. On our own Moon and the moons of Jupiter, features that from a distance look like rosettes are usually impact craters, and I don't see why that shouldn't be the case with the rosette on Rhea."

I walked over to the Von Karman Auditorium—a long, low, glass-fronted building adjacent to JPL's front gate—where six scientists and engineers were about to hold a press conference. There were three hundred or so reporters at JPL for the mission; as they included a number of television reporters, the scientists, normally casually dressed for the most part, were, in many cases, stylishly and impeccably attired (in a downright unscientific manner) in the event, some of them admitted sheepishly, of a summons before the bright lights. The scientists were under other pressures during the mission, for in addition to the daily press conference, there was a meeting of the entire science team (comprising the one hundred and twenty-five scientists for all the Voyager experiments) at eight every morning and another at two every afternoon; a meeting of the imaging team every afternoon at one; plus occasional meetings of the four groups into which the imaging team was divided; as well as meetings of ad hoc committees set up for special purposes (including one that met late every afternoon to pick the pictures to be released at the next morning's press conference). So that it was quite possible, one scientist told me, to spend all day in meetings, without doing any science at all.

While I waited for the press conference to start, I studied a replica of Voyager, identical in nearly all respects to the real thing, which stood to the left of the stage. It weighed about eighteen hundred pounds and consisted of a white, dish-shaped radio antenna twelve feet in diameter atop a ten-sided structure about eighteen inches high and five and a half feet across. The ten-sided structure contained much of the electronic equipment for operating the craft, including its computers and fuel for its sixteen thrusters. The thrusters, for maneuvering, and two star trackers, for navigation, were mounted around the outside. Suspended from one side of the spacecraft on five-foot-long steel-and-titanium struts was its electrical-power source—three radioisotopic thermoelectric generators (RTGs), which were angled away from the spacecraft proper and its sensitive scientific instruments. The scientific instruments were mounted on two booms a third of the way around the spacecraft from each other and from the RTGs. On one boom, forty-three feet long and made of epoxy glass, were magnetometers for studying planetary and interplanetary magnetic fields. The other boom, seven and a half feet long and made of graphite epoxy, held a cosmic-ray detector; plasma detectors for studying, among other things, the solar wind; an instrument for studying low-energy charged particles; and, at the end, a small tray, called the steerable platform, which could turn this way and that. On it were three instruments for analyzing planetary atmosphere, temperature, and surface composition: an ultraviolet spectrometer, an infrared interferometer-spectrometer and radiometer, and a photopolarimeter, which, unfortunately, had broken down during the Jupiter encounter. Also mounted on the steerable platform were two television cameras—one for wide-angle views and one for narrow, high-resolution ones. (The instruments on the steerable platform all had to look at the same thing at the same time, and this had caused a few disputes between the team of scientists that ran them—particularly between the imaging team, whose interests included the rings and the moons as well as the planet, and the infrared team, which needed to spend long periods making thermal mosaics of Saturn.) Two thirty-three-foot antennas trailed off the spacecraft. They were shared by the planetary–radio-astronomy team, which was studying radio signals emitted by Jupiter and Saturn, and the plasma-waves team.

Except for the big white radio dish at the top, the spacecraft was mostly sheathed in black. Generally, spacecraft insulation is white, silver, or gold, to reflect the heat of the Sun, but Voyager was traveling through regions so remote from the Sun that the situation was reversed: black was needed to retain the warmth. At Saturn's distance, shadowed surfaces could drop to a temperature of −320° F. Only the radio dish was white because it would be pointing almost always toward the Earth, and hence in the general direction of the Sun; the rest of the craft, except for the instruments on the booms, would usually be in the dish's shadow. The black sheathing was a multilayered insulation blanket, providing not only thermal control but also protection from micrometeoroids. It was stitched in a loose-fitting way over the spacecraft's odd bumps and angles, and covered the instruments, too—there were holes for lenses, sensors, antennas, thrusters, and star trackers to peep through. It looked like a horse's caparison.

Raymond L. Heacock, the Voyager project manager, opened the briefing by announcing that the spacecraft was operating perfectly at the moment. Moreover, he said, the weather was good at all three of Voyager's signal-receiving stations on Earth—at Madrid, Spain; Canberra, Australia; and Goldstone, California. (Because of the Earth's rotation, three stations, more or less equally spaced around the globe, were required to keep up constant communication with the spacecraft.) Storms would be a concern in the future. If need be, the spacecraft would shift its transmission from what is called the X band, which is susceptible to storm clouds, to the S band, which isn't. There would be a trade-off, though, for while the S band is more reliable than the X band, it transmits at a much slower rate and sends fewer pictures. This is the sort of problem flight engineers seem to face with some regularity.

Then Bradford Smith showed some slides of images that had been taken the day before. The planet glowed ethereally against the black of space—a luminous vision of subtle browns, yellows, and ochers. (Although Voyager's pictures are black and white, in certain cases, where the same picture had been taken several times through different-colored filters, the IPL could put them together into a single color composite.) In one of the pictures, Saturn was seen tilted at an angle, its rings tipped, as if the planet were a child's top at the conclusion of its spin, about to fall over. Because Saturn's equator tilts nearly twenty-seven degrees with respect to its orbital plane, its seasons are very pronounced. They are also very long, since it takes Saturn almost thirty years to go around the Sun. Currently, it was early spring in the northern hemisphere, and it would remain spring there for almost seven years. In the picture, Saturn looked noticeably flatter at the poles than at the equator, and it is: its equatorial diameter is 74,600 miles, its polar diameter only 66,860 miles. This squatness is the result of its rapid rate of spin—once every ten hours, thirty-nine minutes, and twenty-six seconds—combined with its low mean density. Saturn is not a solid planet but a huge ball of gases, principally hydrogen with some helium and smaller amounts of other elements all in roughly the same proportion they are in the Sun. Its mean density—seven-tenths that of water—is less than any other planet's; although its volume is eight hundred and fifteen times the Earth's, its mass is only ninety-five times the Earth's.

On any picture of Saturn, the A, B, and C rings—most readily visible by telescope from the earth—reach so far into space that they dominate, even define the planet, as though its globe were some quintessential definition of planethood, set aside by those glowing parentheses; for while in black and white they look like a curve on a California freeway, in color they are a brilliant whirling wheel of frosty glass that glimmers in the sun. Much, of course, has been learned about the rings in the three centuries following their identification by Huygens, who thought they were a single, solid disc. In 1675, the Franco-Italian astronomer Jean Dominique Cassini found the supposed single ring to be two rings, separated by a gap; this gap is still known as the Cassini division. In 1785, the French astronomer and mathematician Pierre Simon de Laplace declared that there weren't just two rings but a great many, which he characterized as concentric and revolving around the planet. In 1856, the Scottish physicist James Clerk Maxwell deduced that the rings would not be solid but had to be made up of "an indefinite number of unconnected particles"—a deduction that was confirmed in 1895, when it was found that the inner parts of the rings orbited more rapidly than the outer parts. Maxwell's deductions were based on Kepler's laws, which stipulate that objects in lower orbits must revolve faster around a planet than do objects in higher orbits, since the lower the orbit is, the greater is the velocity required to overcome the planet's gravitational pull. (This, of course, was why Smith had said the spokes, which run ten thousand miles radially across the B ring, shouldn't be there: they should have been sheared apart by the varying speeds of the orbiting particles within the ring.) Maxwell's "indefinite number" of ring particles might better have been an "infinite number," for each ring is composed of trillions upon trillions of particles, presumably ranging from very small to very large (how small and how large, and what was their typical size in each ring, were some of the things Voyager would try to find out). Long before Voyager, Earth-based spectroscopy had found that the surfaces of the ring particles consist of water ice, and that accounts in part for their dazzle; though at the time it was thought the water ice might coat particles that were basically silica, later analysis of ground-based radio and radar observations suggested that the particles consist almost entirely of water ice. (There is no danger of the rings' or the moons' melting, for the Sun's energy in the vicinity of Saturn is one percent of what it is at the Earth.) The rings are extremely thin—at the time, they were thought to be perhaps a mile thick, though it was recognized that they could well be much less than that—and their thinness enhances their appearance of icy delicacy. When they are seen edge-on from Earth—something that happens approximately every fifteen years—they appear to have a crystallike sharpness.

Of the rings that are most visible from the Earth, the innermost is the C ring, which begins about eight thousand miles above Saturn's cloud tops and is about twelve thousand miles wide. Much dimmer than the A and B rings, it has reminded some observers of a flimsy, rippled piece of paper; it is often called the crepe ring. (The differences in the appearances of the rings, some scientists guessed, had to do with differences in the size and number of particles.) Next comes the B ring (the rings were assigned letters according to their dates of discovery, not their positions), which adjoins the C ring and is sixteen thousand miles wide. The Cassini division, some two thousand miles wide, separates the B ring from the A ring, which is ten thousand miles wide. The A ring has a small interior gap two hundred miles wide named the Encke division, but generally referred to as the big A-ring gap. These are the three main rings, but several more were discovered in recent years by telescope and by Pioneer 11. The existence of one, the D ring, which some astronomers had seen—or thought they had seen—from the Earth in 1969, had yet to be confirmed. It was believed to be a faint miasma adjoining the inner edge of the C ring and extending down toward the surface of Saturn; there was some thought that its particles fell continually on the planet's equator. Another, the F ring, was definitely discovered by Pioneer; it is an elegant stringlike circlet situated some twenty-seven hundred miles beyond the outer edge of the A ring—in the Saturnian scale, practically next door. There had been rumors of another ring, lying some eighteen thousand miles beyond the F ring; it was too thin to be seen by Pioneer's optics, but it appeared as a dip on a graph transmitted by another of Pioneer's instruments, which measured charged particles. Finally, there is the E ring, a toroidal—doughnut-shaped—haze that begins about ninety-three thousand miles above Saturn's cloud tops (far outside the other rings and just outside the orbit of Mimas), reaches its maximum intensity at the orbit of Enceladus, some hundred and twelve thousand miles out, and extends more faintly out to the orbit of Rhea, some two hundred and fifty thousand miles out. It was first seen from the Earth, in 1966, but it wasn't named until after the D ring because its existence, too, had been challenged.

Smith next displayed a high-resolution Voyager photograph, taken the day before, of a section of the rings, showing clearly something that Pioneer's less powerful optics had only hinted at—that the three major rings, A, B, and C, were themselves apparently divided by tiny gaps (the "tire marks" I had seen in the raws) into a myriad of smaller rings, and that the Cassini division contained little rings of its own. "It's a good bet there's even more structure in there," Smith said. "So we are looking forward to seeing closer-up photographs in the days ahead." Next he

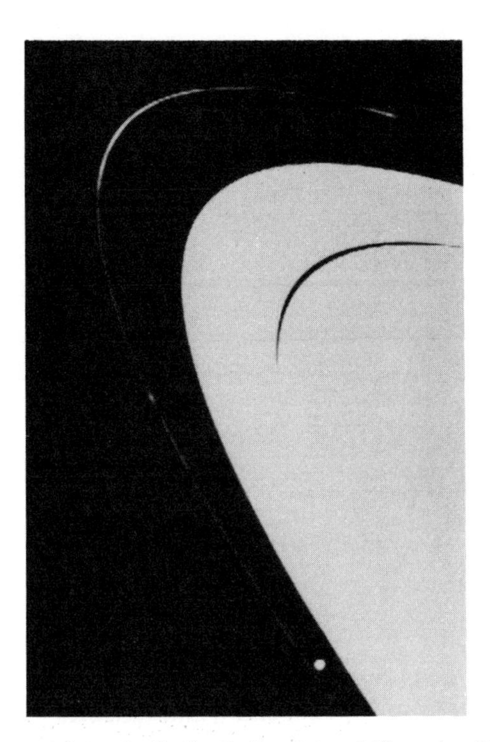

S14, just inside the F ring. Imaged November 8, 1980.

showed a picture of the rings taken with an ultraviolet filter; the particles in the C ring and within the Cassini division appeared blue, and the other rings showed up as yellowish. The blue color, Smith said, indicated that the C-ring and Cassini particles were somehow different—perhaps in composition, in configuration, or in size—from those in the A and B rings. Then came a picture of the F ring, which was still a very thin string—less than fifty miles across—but now appeared to have small clumps here and there. "We have no explanation of what causes those clumps, what holds them together, or how long they last," Smith said.

Smith pointed out a white dot just inside the F ring; it was one of a pair of tiny moons flanking that ring. These moons, discovered by Voyager a few weeks earlier, had been named, for the time being, S13 and S14. S13's orbit was quite eccentric, ranging from about three hundred miles to twelve hundred miles outside the F ring, and the orbit of its partner, S14, was about three hundred miles inside it. Both moons were about fifty miles in diameter. A third new moon, S15, had been discovered only three days earlier. It lay just outside the A ring and was even tinier. Smith said that the gravitational force of these three moons evi-

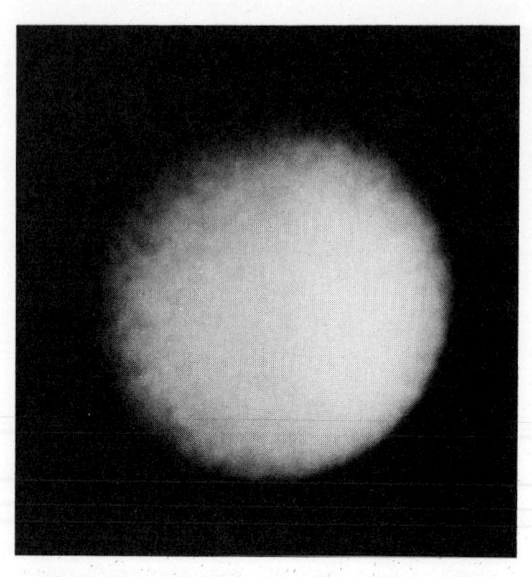

Titan, a fuzzy-looking cue ball, enticing even from seven billion five hundred and sixty million miles away. Imaged November 4, 1980.

dently kept the particles within the neighboring rings from straying. S13 and S15, on the outer boundaries of the F and A rings, and (according to Kepler's laws) moving more slowly than the particles inside them, would exert a gravitational drag on any particles moving outward and cause them to fall into a lower orbit; S14, on the inner side of the F ring and moving faster, would speed up the particles it passed, boosting them into a higher orbit. In this manner, the ring boundaries would be maintained, and the three moons were therefore being called "shepherding" moons. (This notion had first been suggested by a couple of astrophysicists, Peter Goldreich, of the California Institute of Technology, and Scott Tremaine, of the Massachusetts Institute of Technology, to explain the nine sharply defined rings of Uranus, whose moonlets, if there are any, are far too small to be seen from Earth.) Some scientists thought that the boundaries of other rings would be good places to look for still other moons.

The Voyager scientists also hoped to learn more about two moons, S10 and S11, that had been discovered by ground-based observation (Pioneer had also verified one of them). They circled about seven thousand miles outside the F ring. Both of them had odd shapes (S11 looked like a molar tooth), and both were small, but still about twice the size of the F-ring moons. They were in almost identical orbits, only thirty miles apart. The existence of these two co-orbitals, as they were called, had generated a mystery, because the moon in the inner orbit should move faster than the moon in the outer orbit and periodically overtake it, the theoretical result being a celestial crackup—but, clearly, no such periodic crackup was occurring. Scientists now think what may be happening is that, as the inner moon begins catching up with the outer moon, it begins to feel the latter's gravitational pull. Thus energized, the inner moon rises to a higher orbit and then slows down. Meanwhile, the outer moon has lost energy and dropped to a lower orbit; now this moon is traveling faster—thus the moons exchange orbits and speeds.

Finally, Smith showed a slide of Titan, Saturn's biggest moon—a lovely opalescent orange, shimmering delicately against the black behind it. Clearly, if Saturn and its moons were part of some sort of celestial billiard game—a favorite metaphor of the press—then this was the cue ball. In the coming days, Smith said, Titan would be the ball to keep one's eye on.

Back in the imaging team's headquarters, Reta Beebe, a dark-haired, lively woman in her early forties, was in one of the interactive rooms, working her way through pictures of Saturn taken shortly after nine that morning, when Voyager was two million nine hundred thousand miles away from the planet. Beebe, who is an astronomer at New Mexico State University, was searching the pictures for identifiable features on Saturn's disc. (A planet's disc is the seemingly two-dimensional aspect of its globe as seen from a single, distant perspective.) Like Jupiter, Saturn has several lateral bands across its disc, but whereas the bands of Jupiter had been pronounced, Saturn's bands seemed faint. Certain of the brownish markings on both Saturn and Jupiter are thought to be organic compounds. All in all, Saturn's disc seemed something of a washout compared to Jupiter's—the word the astronomers used to describe it was "bland." Jupiter, a sixth larger in diameter than Saturn but twice as dense and over three times as massive, was an angry-looking planet with a vast, roiling welt, the Great Red Spot, seething in the southern hemisphere. Three other spots, among them the White Oval, seethe south of it; and with the forceful bands of color paralleling the equator—reds, oranges, browns, and whites—they looked as if they had been laid on in a fit of temper by Van Gogh; whereas Saturn, apparently more delicate and luminous, might have been painted by Monet on a misty morning. (The same difference in intensity extended to the rest of Jupiter's system, in comparison with Saturn's: four of its moons, Ganymede, Europa,

Callisto, and Io, called the Galilean satellites because they were discovered by Galileo, were of the same large size as Titan, Saturn's single giant satellite; and one of them, Io, whose surface was a fiery red, was shooting sulfuric plumes a hundred miles into the air, in a manner not anticipated in the environs of Saturn. The only delicate thing about Jupiter was a thin ring discovered by Voyager.) On Saturn, there were a myriad of small spots, fainter than the ones on Jupiter, Beebe said, which apparently moved with the winds, and were useful for determining their speed and direction. Plotting the movement of such features could ascertain planetary meteorology. However, most of the small spots kept disappearing. For this reason, Saturn's wind patterns were proving difficult to establish; nevertheless, Beebe felt she was beginning to get a sense of them. "At the equator, the flow seems to be eastward; above it and below it, there are bands where the flow is westward," she told me. "We know very little about Saturn yet. The upper part of the atmosphere, we now think, is probably a haze of hydrogen and helium gas containing particles of other compounds—it's like Jupiter's but optically thicker. At some lower depth, perhaps fifty miles down, there are clouds of frozen ammonia, and these are even more opaque; they seem to reflect back—backscatter—sunlight. So that what we see, we think, is basically light reflected from that ammonia cloud deck, plus the haze above it, which is backlit." She pointed to some brownish-yellow shadings on one of the pictures. "The colors are from the ammonia cloud deck—they're determined by the absorption of sunlight there. But the haze seems to be greatly reducing the contrast. Essentially, the same situation exists on Jupiter. We think that one reason there is less contrast on Saturn is that Saturn is colder, so the frozen-ammonia cloud deck forms deeper down on the planet—and that means that the haze above it is thicker."

At the time of the press conference next morning—10:00 A.M. on Monday, November 10th, one day before the encounter with Titan—the spacecraft, having sped up now to about thirty-five thousand five hundred miles per hour, was almost two million miles from Saturn, having traveled some nine hundred thousand miles in the previous day. Most of the time, the spacecraft flew motionless through space, its big white radio antenna on top pointing back toward the Earth, kept there by occasional bursts of its thruster jets, two pairs each, facing in opposite directions, for the three major axes of motion: roll, pitch, and yaw. (The spacecraft has two redundant sets of these, making twelve pairs of thruster jets, plus four more for trajectory changes, or a total of sixteen pairs; they operate on a hydrogen-nitrogen gas called hydrazine.) The thrusters are controlled by the attitude and articulation control subsystem (AACS), which in turn receives much of its navigational information from the Sun sensor, which—because the antenna, usually facing the Earth, is also facing the Sun—peers through a hole in the antenna dish in order to control two of the major axes of motion, pitch and yaw; and also from the single star tracker (there is a backup for it), which recognizes the light from the brightest stars in a ribbon all around the heavens—about a dozen were used as guide stars during the mission—in order to control the third and most used axis of motion, roll. The night before, the spacecraft had changed to a new orientation or attitude, which it would hold, with occasional movements for better viewing, for the rest of the inbound leg toward Saturn, in order to be in the best overall position for looking at the various rings and moons. For each attitude change, the thrusters spat out two identical jets of cold hydrazine gas—one to start and one to stop. Rolling is the easiest of the three orientation maneuvers because it can roll without losing radio contact with the Earth; the star tracker, which had been fixed on Canopus, shifted to Miaplacidus. (In the billion-mile trip, the spacecraft, with one sensor always aimed at the Sun, had made a sweeping outward arc, so that some of the guide stars had changed position in the sky, in relation to the star map loaded into Voyager's computer before it left Earth—the map's reading for Canopus was off by as much as fifteen degrees. As the computer's map couldn't be changed, the star trackers were biased electronically, so that *they* did the shifting, to allow for the outdated map.) The spacecraft paused during the roll maneuver in order to do an entire rotation around its yaw axis, sweeping three hundred and sixty degrees around Saturn's equatorial plane and returning the Sun sensor to the Sun; the scientists for some of the instruments fixed immovably to the spacecraft (in particular those instruments studying electrical and magnetic fields or atomic particles emanating from Saturn) wanted to sweep the sky in that plane. (The maneuver also enabled those instruments to look at fields or particles emanating from the Sun, for when the spacecraft flew by, the Sun—and also the Earth—were in the plane of Saturn's rings.) "Normally, we don't want to make maneuvers like this so close to encounter," Richard P. Laeser, the mission director, told me on the way in to the press conference. "If there is trouble, and the spacecraft loses contact with the Earth, it might be ten or twelve hours before it can find the Earth again. The computers aboard are programmed to do this, but there could be a tremendous loss of data before

Dione, looking "ropy" or "wispy," as Schiaparelli might have seen it. Imaged November 9, 1980.

it does." With the sensor safely back on the Sun, the spacecraft completed its roll until the star tracker locked on Miaplacidus.

At the press conference, Smith showed some pictures taken the previous day. In one, Saturn looked yellowish-gold, with the rings a handsome beige; it drew a gasp and some applause from the audience. A shot of Titan now showed it to be an even more brilliant orange. Smith, who was giving a running commentary, pointed out a line that had been discovered by Tobias Owen, the head of the imaging team's Titan group. The line seemed to run along Titan's equator; indeed, the satellite's atmosphere north of the line appeared to be a somewhat darker orange than the atmosphere to the south. (Titan was becoming increasingly interesting; William Sandel, of the ultraviolet-spectroscopy team, had previously announced that his instrument had discovered Titan's atmosphere was apparently leaking hydrogen into space, for Titan appeared to be traveling around Saturn inside a huge toroidal cloud of that gas, apparently its own emanation.) Smith also showed the picture of Rhea that had impressed Masursky on the interactive the day before;

it, too, looked more striking now that it had been through the processing lab. Pointing to the hazy rosette—yesterday's dim spot—Smith, as Masursky had done, said he was tempted to think of it as a meteoric impact. Next came a slide of the moon Dione, whose diameter is about a third of our own Moon's. The picture showed what looked like ghostly smudges here and there. "We are definitely seeing things now on the moons—although we're not sure what they are," Smith said. The rosette on Rhea, and particularly the smudges on Dione—large, sprawling mottlings, which members of the imaging team referred to variously as "wispy" or "ropy"—resembled the Rorschachlike spots on drawings of Mars by nineteenth-century planetary astronomers such as Eugène Antoniadi, Giovanni Schiaparelli, and Percival Lowell, who had tried to record accurately what they saw at the limit of the resolution of their instruments. According to Laurence Soderblom, Smith's deputy team leader—a geologist in his late thirties with light hair and a long, thin face—as powers of resolution grew greater, it seemed that before the planets could be seen crisply and clearly, they passed through a stage in which their various features could be seen only as splotches, and it was through that stage that Voyager was passing now. If Lowell saw Mars the way we were seeing these moons, then it's little wonder he thought he saw canals.

After the press conference, I visited Smith in his office and asked him how he had become involved with space missions.

"Growing up within biking range of the Harvard observatory helped," he said. "When I was in my teens, Leon Campbell, an astronomer there, got me interested in studying variable stars—plotting their light curves was something an amateur could do." Like many a young would-be astronomer, Smith did not realize that astronomy was a field in which he might actually make a living, so he majored in chemical engineering at Northeastern University, in Boston. After he graduated, in 1954, he was drafted, and the Army sent him to work at the New Mexico College of Agriculture and Mechanic Arts, in Las Cruces, New Mexico, where Clyde Tombaugh, the astronomer who discovered Pluto in 1930, had undertaken a search for tiny moonlets that he thought might lie between the Earth and the Moon. The Army was interested in using such moonlets for the triangular measurement of intercontinental distance on Earth as an aid to targeting ballistic missiles. Tombaugh and Smith set up an observatory at the college. No moonlets were found (apparently none exist), but Smith so fell in love with astronomy at Las Cruces, where the nights are cold and clear, that when he got out of the Army he stayed on at the observatory. There, with Tombaugh, he set up a program for watching Venus, Saturn, and Mars continuously. Unlike many traditional astronomers, Smith was coming to view the planets as dynamic, changing bodies. "We began to see seasonal changes on Jupiter," he said. "We detected the vorticity of Jupiter's Great Red Spot—we learned that it was a giant anticyclone. We studied Saturn in the same way. It's hard to remember now what it was like to start out, as I did, in the nineteen-fifties, viewing planets telescopically and wondering whether anyone would ever go there. I think that I'm probably more emotional about Voyager than many of my younger colleagues, because I spent so many years peering at these places through telescopes, struggling to see them only just a little bit better—and here we are, seeing them close up." In 1964, Smith was appointed director of the observatory at the college, which by this time had become New Mexico State University. Ten years later, he took over the planetary-photography program at the University of Arizona. His predecessor in that job had been Gerard Kuiper, a courtly Dutchman who was then regarded as the dean of planetary astronomers. Kuiper had made his reputation in that field in the thirties, when such studies were at a low ebb. (This was partly because of the excesses of Percival Lowell, the founder of the Lowell Observatory, who made enormous claims for his Martian canals, and partly because of the immense popularity of the new field of astrophysics.) Kuiper and Tombaugh, in fact, were a bridge between an earlier era of planetary astronomy and the space age. It was Kuiper who, in the 1940s, discovered that Titan has an atmosphere—an atmosphere, moreover, containing methane. Two of the Titan specialists on the Voyager imaging team, Tobias Owen and Carl Sagan, had been students of Kuiper's.

Smith had first become involved with spacecraft with Mariner 4's flyby of Mars in 1965, at the behest of the man many consider the founder of planetary astronomy by spacecraft—Robert Leighton. An expert in television optics, Leighton put together the imaging team for Mariner 4. (This mission would discover that Mars, far from being covered with greenery, as many still believed in a sort of holdover from Lowell, was cratered and apparently dead.) Aside from Smith, who himself was rapidly becoming an expert on electronic observation, Leighton also involved on that team Bruce Murray, then a professor of geology at the California Institute of Technology and currently director of JPL; Harold Masursky, of the astrogeological branch of the United States Geological Survey (USGS); and Merton Davies, a geodesist with the Rand Corporation; all three would later be involved with Voyager. (One thing about planetary missions is that the same faces keep popping up over and over again, in different combinations: Smith, Murray, and Davies went on to Mariners 6 and 7, two more craft that flew by Mars in 1969, and thence to Mariner 9, which went into orbit around Mars in 1971. Masursky, who had been associated with various Ranger, Surveyor, and Apollo missions to the Moon, was the leader of Mariner 9's imaging team, and Smith was his deputy. Mariner 9 also included Sagan from Cornell, who went on to Viking, to the Mars landing mission, and then to Voyager; and Soderblom, who had recently received his doctorate from the California Institute of Technology, where he had been a student of Murray's, and who later went on to USGS's astrogeological branch.)

In the late 1960s, preliminary plans were getting under way at NASA for a mission that would exploit the alignment in the early 1980s of four of the five outer planets—Jupiter, Saturn, Uranus, and Neptune—by taking advantage of the gravity of each to swing a spacecraft to the next. (Tombaugh's Pluto was not part of the lineup.) "Such an alignment occurs once every hundred and seventy-six years," Smith told me. "The last time was during the administration of Thomas Jefferson, and the wags around here are saying that he blew it." NASA almost blew it, too, by proposing a too elaborate mission, called the Grand Tour, which was rejected by Congress. The less ambitious Voyager was approved in 1972.

All in all, Voyager's imaging team was a formidable group of veteran planetary troupers. Originally made up of eleven people (including Smith, Davies, Owen, and Masursky), it picked up nine more associates, nonveterans mostly, before the spacecraft reached Jupiter, and six more—some of them experts in rings—were added before the Saturn encounter, making a total of twenty-six. The newer members, though, were by and large very much out of the same mold as the older ones—from the Geological Survey's Branch of Astrogeologic Studies, from the University of Arizona, from the California Institute of Technology, or (in a couple of cases) from Cornell: they tended to be associates of the original members or their past students or the students of students, just as some of the original members had been of each other. Most of the team members were in their thirties and forties, with Masursky and Eugene Shoemaker, another senior geologist with the astrogeological branch, both in their fifties, being among the oldest; there were some in their twenties, too. Some of them had been working together on Voyager for eight years (longer, if you count earlier missions); already veterans of two encounters with Jupiter, they all knew each other very well and, despite some rivalries, got along easily—often the case with people allied in a project that excites them.

The structure in the Cassini division that separates the A ring (left) from the B ring (right). Imaged November 8, 1980.

I joined a group of several scientists who were sipping coffee in one of the interactive rooms, watching a scientist flipping through the latest complete disc pack of pictures taken during the night. One of the people standing around was Tobias Owen, Kuiper's student, a dark-haired scientist in his forties who is now an astronomer at the State University of New York at Stony Brook and an expert on planetary atmospheres. "That's the Cassini division," he said, putting down his coffee and leaning in front of him to get a better look. "It's all filled with rings, we see now, but there are still distinct boundaries between it and the A ring and the B ring. I'm trying to see if there might be a couple of moons, one on each boundary of the division, 'shepherding' the rings on each side, as is the case with the F ring. The boundaries are so sharp, there might be." If there were, however, they weren't apparent. The material gave off about as much light as that in the relatively dim C ring, so it wasn't surprising that it had not been spotted before—from Earth, its rings were invisible because the division was squeezed so closely between the A and B rings' brilliance. The Cassini division was so full of structure—like the C ring, the material resembled concentric segments of thin bicycle tires, tubular but ghostly—that, one scientist commented, it was lucky Pioneer 11 had not been sent through it, as had once been contemplated, or that mission would have ended right there. (The reason it hadn't been was that Smith, who had noticed a blurring in the Cassini division on some high-resolution photographs taken from Earth and guessed that it contained some material, had made a big fuss.) The picture on the screen changed quickly through a series of several shots of different parts of Saturn taken with an orange filter on the lens, which would bring out any methane in the atmosphere, and several shots of the moon Tethys, which was still too far away to be more than a blurry white ball. "It's almost as bad as a Pioneer image," the man at the controls said, getting up. (The Voyager scientists often seemed to put down Pioneer—even though it had gotten to Saturn first, it had gotten there with the least.) "Anyone else want to play? Anyone else want to fly the spacecraft?"

Owen sat at the controls, which at first he had trouble mastering. Evidently he was not good at machinery. "I usually try to get somebody else to do it," he said. He produced a new picture of Titan and then studied it to see if it would show the equatorial line and the relative darkness of the northern hemisphere that Smith had reported Owen had noticed earlier. At first, he didn't enhance it enough. "Sometimes it helps to look at these things at three o'clock in the morning, after several cocktails," he said. At last he produced the atmospheric effect. "I'm beginning to think that the darkening might have to do with the magnetic field of Saturn," he said. The magnetic lines of force, emanating from the two poles of Saturn and looping hundreds of thousands of miles out into space until they met, forming a huge belt around Saturn, would hit Titan from above and below. "The magnetic field might affect the atmosphere in the northern hemisphere differently from the atmosphere in the south—I don't know why it should, but it's the basis of an idea. It's the only mechanism I can think of that would divide the two hemispheres so perfectly." Owen dropped the idea a day or so later when a magnetic expert told him there was no way Saturn's field could cause the color difference. Right or wrong, the scientists were being provoked by the new data to come up with explanations for what they were seeing—and some new ideas were being tested against the data, too. One such was that Titan, the size of Mercury, might have a small moon of its own. Owen began looking around Titan's environs to see if he could find it. He couldn't. The looking would be better after tomorrow night, he said, when Voyager had passed Titan and, the Sun behind it then, the satellite and any mini-moons around it would be back-lit, making any such moonlets (a crescent moonlet, it would be) easier to see.

In the second interactive room, Jeffrey Cuzzi, a tall, thin, young planetary scientist from NASA's Ames Research Center in Mountain View, California, was enhancing a picture of the rings. When he finished fiddling with the picture, the rings appeared to consist of even more ringlets than they had the day before, with narrow divisions between them. "Sometimes it's not clear whether we're looking at rings or divisions," Cuzzi said.

The discovery of an increasing number of divisions in the rings was of interest to the scientists in part because of a theory that such divisions are caused by what are called resonances with the big moons of Saturn, beyond the rings. Much of the work on resonances had been done by three astronomers at the Smithsonian Astrophysical Observatory, in Cambridge, Massachusetts—Fred A. Franklin, Giuseppe Colombo, and Allan F. Cook II. Two of them were at JPL for the Voyager encounter: Cook was on Voyager's imaging team, and Franklin was there as a guest observer. As Cuzzi explained it, the big moons, which according to Kepler's laws circle Saturn less often than the rings, created within the rings what were called "resonance regions." These were places where the orbital period of a ring particle would be a simple fraction—say exactly a half, or a third, or a fourth—of that of one of the moons. Whenever *any* particle, on its inside track, caught up to a moon, on its outside track, the moon's gravity gave it a little push; if in successive passes the push came at different points in the particle's circuit (as would be the case in nonresonance regions), the effect would be too random for anything to happen. However, if this happened to a particle at the same point during each orbit (as would be the case in the resonance regions) the cumulative effect would cause the particle's orbit to become somewhat elongated. It would thus collide with neighboring particles, and a narrow division—a gap—would result. Cuzzi likened resonances to a man pushing a swing:

if he gives a little shove at just the right moment—the top of each backward sweep—he can be said to be in resonance with the swing's motion, and the swing will go higher than it would if he pushed it at any other point in the swing's arc. Mimas, the innermost of the large moons, was of particular interest to the resonance theorists, because its mass and its proximity to the rings meant that it would have a substantial effect on them. Ring particles that made two complete circuits for every one that Mimas made were said to have a two-to-one resonance with it, those that orbited three times a three-to-one resonance, and so forth.

Fred Franklin, who was in the room and happened to hear our conversation, came over and said, "We've been trying to fit some of the new gaps with other moons. And it may be that we can use the gaps to find some more moons."

Cuzzi nodded. "The trouble is, Voyager is still a long way out," he said. "But we're already finding more gaps than we can account for by the resonance theory. And some of the gaps are wider than they ought to be; according to the resonance theory, they should be quite narrow. We certainly *are* finding gaps consistent with the resonance theory, but there seem to be a great many others as well, so something else may be going on."

I asked him how he planned to tackle this problem, and other problems, in view of the stream of data that was already coming in and that was bound to increase.

"Normally, when you confront a new set of data the first thing you do is look at the theories you already have, such as the resonance theory—you see if they fit, if they explain the situation. Next, you look for anything that resembles something you have seen before. For example, we have seen the moonlets—S-Thirteen, S-Fourteen, and S-Fifteen—shepherding the F ring and the A ring, and we are fairly sure that moonlets can sharply define the orbits of particles. Perhaps there are more moonlets embedded in the rings, shepherding them from there. The third thing you do is to think of any new physics or new theories that might apply. We haven't got to that stage yet, but we may soon—to explain the spokes in the B ring, for instance. Finally, you go through the data slowly, classifying everything, building up a body of facts that can be correlated and that might lead you to new theories. This last phase will probably take a decade or so. But right now, with the data still coming in, we mostly end up just sitting around looking at the interactives and batting ideas back and forth. We're like a group of people on a beach with waves breaking over us; each one of us picks up whatever he can get hold of—a shell, a starfish, a piece of glass—to look at. Right now, there's no particular order to anything—though Fred, who has been working with resonances, naturally tends to focus on them. Often, young scientists who have not yet developed a specialty are freer to pick up totally new things."

Franklin nodded, and said, "Imagine having all your information increasing a thousandfold in a matter of hours! It's as if, in the arts, someone came up with a whole treasure trove of new Shakespeare plays."

A short, elderly man with thin white hair was silently studying some new photographs of the rings that had been taped to the wall outside one of the interactive rooms. He turned out to be Clyde Tombaugh, the discoverer of Pluto—a guest of his onetime assistant, Bradford Smith. I asked him how he felt about astronomy by spacecraft.

"It certainly beats staying up all night in a cold observatory dome," he said. "Spacecraft are handy. However, I'm of the old school—I believe that you should do it the hard way. We really need both types of observation, though, and I'm delighted with what Voyager is getting—so much more detailed than anything we can get from the ground. My only regret is that Voyager Two won't be going on to Pluto. Pluto, unfortunately, won't be in the right place at the right time."

Several scientists had gathered around us to look at a ring picture that had just come off one of the interactives as a hard copy. The rings were well lit, the Sun being more or less behind the spacecraft and hitting the rings at a three-degree angle. (Before the encounter period was over, the scientists would have two other views of the rings: from underneath, after the spacecraft passed below the ring plane, and from above once more, when it was on the outbound leg and was looking back toward the Sun.) Some of the scientists laughed in disbelief as they saw the huge number of narrow ringlets that the rings had apparently divided into, which exceeded anything they had imagined; something like a hundred had been counted a day or so before, but now the count looked closer to five hundred. There were so many gaps and ringlets, one scientist said, that the whirling disc looked like a phonograph record. "It must be where the music of the spheres comes from," another scientist said.

I asked Bradford Smith, who had in a spare moment joined the group looking at the picture of the rings, a question that had been bothering me ever since I had heard of the newly discovered moons inside the F ring—the A ring's shepherd and the inner shepherd of the F ring—and that bothered me even more now that there was talk of additional even smaller moons embedded inside the broad A, B, and C rings: Rings and moons, I had been told during discussions of the origins of our own Moon, are not supposed to mix, because there is a sort of invisible barrier around any planet, called the Roche limit, a few radii away from its surface, inside which any moon that happened to wander would be torn apart by the planet's tidal forces, the pieces perhaps forming a ring around the planet; conversely, any ring of particles *outside* the Roche limit would immediately clump together to form a moon. What were the two new moons doing inside the F ring? What was the F ring doing outside the two moons? What were the scientists doing looking for moonlets embedded in rings even nearer Saturn? Smith explained patiently that there was nothing magical or arbitrary about a Roche limit; as it depended on the difference between a planet's tidal pull on the near and far sides of any moon, the point at which a moon broke up would depend on the moon's diameter, as well as upon the material of which it was composed. The smaller and tougher a moon, the nearer in it would survive; the known new inner moons were evidently small enough and tough enough (the thinking was that they, like the larger moons, were made predominantly of ice) to survive so close to Saturn. Any hypothetical satellites embedded in the inner rings would be made of ice also and be even smaller—a few kilometers, or tens of kilometers, across. Smith pointed out that the difference between the biggest ring particles and the smallest of the new moonlets that the scientists hoped to find embedded in the rings could simply be just a matter of degree. An embedded moonlet might never be seen, he added, because Voyager 1 would not get close enough to the rings to make out any object smaller than about five miles across. The ring particles—ranging in size, no doubt, from icebergs to boulders to pebbles and even dust—probably varied from ring to ring and, for that matter, might vary considerably within each ring. More would be known when Voyager went beneath the rings and then beyond them, because studying them from different vantage points would tell a good deal about the size of the particles. For example, while Voyager was beneath the rings, the radio antenna would beam a signal to the Earth through the ring plane, and this would provide details about the size of the particles. Whatever their size, their shape was thought to be angular, the result of several billion years of bumping and jostling each other.

There had been considerable speculation about what such whirling sheets of independently orbiting pieces of ice were doing spinning so smoothly around Saturn in the first place. Smith and others had long been puzzled by the fact that the rings were there at all, for rings of particles, by their very nature, ought to be unstable. That mystery was partly solved, of course, by Voyager's discovery of the three shepherding moons. Without them, the thinking was, all the rings might have been like the hazy E ring, which was so spread out that it could be seen from Earth only when it was edge-on. "The rings are probably four and a half billion years old—as old as Saturn—and they will probably last as long as the solar system itself; that is, another five or six billion years," Smith had said at one press conference.

As to how they got there, either the rings are the remains of a moon that once wandered too close to Saturn and broke up (or was smashed to pieces by other means) or they represent primordial material that was left over from the formation of the planet and never formed a moon. If all the material in the A, B, C, and D rings were rolled together, it has been calculated, it would make a satellite about the size of Mimas—a calculation that could support the notion that a Mimas-size moon had broken up close to Saturn. However, many Voyager scientists think that such a moon would not have shattered into so many tiny fragments—it would have been more apt to break into a few big ones. They seem to find the idea that the rings are a remnant of the formation of the planet more appealing anyway, for then the entire Saturn system—planet, moons, rings, and any moonlets inside the rings—could perhaps be connected in a single theory, and simplicity is something scientists always like to achieve.

Later, I asked Cuzzi to explain the theory that the rings might be a primitive remnant of the formation of Saturn.

"We have to go back to the solar system, which was formed in somewhat the same way," he said. He explained that the solar system developed, about 4.6 billion years ago, from a vast rotating cloud of hydrogen gas (called the solar nebula) enriched with small amounts of heavier material left over from the explosions of distant supernovas. The spherical cloud—originally of a diameter far greater than the present diameter of the entire solar system—condensed by means of its own gravity in such a way that the material at the center coalesced into the Sun, whose nuclear energy caught fire when it reached a certain density and temperature. The remainder of the nebula settled into a wide spinning disc, reaching at least beyond Neptune's present orbit, five and a half billion miles in diameter; a cloud spinning around a central core has a tendency to flatten out in this fashion. "It is a phenomenon we see quite often—instances range from galactic discs to the rings of Saturn," Cuzzi said. First the material in the disc was sorted by the heat at the center, a heat far greater than the Sun's today: volatiles (materials that vaporize easily, such as hydrogen, helium, and water, a compound abundant in the universe) were driven to the cooler outer reaches of the solar system, to the future orbits of Jupiter and Saturn, and beyond. Heavier, more refractory elements (those that don't vaporize easily, such as most rocks and metals) remained behind, so that today the inner planets are composed mainly of those materials and are not enveloped in huge hydrogen atmospheres. Nonvolatile material, of course, was present in the orbits of the outer planets, too, though they were overwhelmed by the superabundance of hydrogen and helium (some of which properly belongs to the inner solar system); Saturn has today, for all its giant gaseous girth, a rocky or metallic core only about double the size of the Earth.

In a process that is by no means completely understood, the material began to accrete in various regions around the newly forming Sun, clumping together into bodies, perhaps a mile wide, called planetesimals, which finally combined to make the four inner planets and the cores of the outer planets. Resonances may have had something to do with the original clumping of the planets, and that is one reason that scientists are interested in studying resonances in the rings of Saturn. The succession of planetesimal impacts created considerable heat on the forming planets, which, like the Sun itself in its earlier stages, were surrounded by huge nebulae. Jupiter, Saturn, Uranus, and Neptune became massive enough to acquire dense envelopes of gases—principally hydrogen—around their cores. (Pluto, a solid rocky or icy body not shrouded in hydrogen despite the fact that it is the outermost planet, is thought by some astronomers to be a satellite that escaped from Neptune—whose orbit Pluto's intersects.) Indeed, Jupiter and Saturn, with their hydrogen envelopes, are sometimes thought of as mini-suns, too small to light up. Like the Sun, Jupiter and Saturn might in their early stage have been several hundred times their present size, though they would soon have collapsed rapidly to just several times their present size. (Saturn probably filled a volume three hundred and seventy thousand miles in diameter, or out to the present orbit of Tethys.) After that, they continued to collapse, but more slowly.

According to another member of the imaging team, James B. Pollack, who is a colleague of Cuzzi's at Ames and who had helped develop many of these ideas in papers published in the mid-1970s, it may have been during the period of initial collapse that the nebula of gases and other materials settled into a disc around Saturn, from which the planet's satellites and surviving rings formed. In many respects, the situation would then have been a replay of the formation of the solar system. As Saturn continued its contraction, moons accreted around it. Cuzzi's best guess about what created the icy Saturnian ring system is a theory that he credits to Pollack: Jupiter and Saturn had managed to attract, as part of their huge nebulae, a lot of water, which, because of the heat emitted as they collapsed, stayed in the form of vapor. Because Saturn, being less massive than Jupiter, cooled more quickly, Saturn's surrounding vapor cloud froze into rings while Jupiter's remained vapor. According to Pollack's theory, at about this time the Sun may have passed through what is known as the T-Tauri phase—a period in the life of a young star when it generates a violent solar wind, electrons streaming away from the Sun. The solar wind would have blown away all the material in the newly forming solar system that hadn't gone into the making of a moon or a planet or a planetary atmosphere, or that hadn't solidified into particles. Saturn's ring system, having crystallized before the T-Tauri phase, would have remained unaffected by it; the planet retained its rings. Jupiter's water vapor, of course, would have been blown away. (Though Jupiter and Uranus have ring systems, recently discovered, and Neptune may have, none is remotely as grand as Saturn's, and none appears to contain water ice, which make Saturn's rings so bright.)

I asked Cuzzi whether cosmologists—scientists who study the formation of galaxies, stars, and solar systems—had shown much interest in Voyager's imaging of Saturn's rings.

"They haven't had time to yet," he said. "But they will."

The next morning—Tuesday, November 11th—the sign in the middle of JPL had been changed to indicate that the spacecraft's distance from Saturn was one million sixty-six thousand miles; that the spacecraft was approaching Saturn at 36,947 miles; and that there was only one day to go until the closest approach to the planet. (Voyager 2, on the other hand, was seven hundred and eighty-four million miles from the planet and had two hundred and seventy-nine days to go.) This was the day of the encounter with Titan. I arrived at the morning press conference in time to hear a reporter ask whether there was any hope of there being a rift in Titan's cloud cover enabling anyone to see the satellite's surface, and Brad Smith's reply, "I have seen nothing to give me an iota of hope." Another reporter asked whether, in retrospect, the scientists were sorry now that they had programmed Voyager 1 around Titan, with the other moons being fitted in as best they could, and with the spacecraft later being shot out of the plane of the ecliptic. Edward Stone, the chief scientist—an angular, sharp-faced physicist from the California Institute of Technology—answered firmly, "No. Even if we can't image the surface, it's the only moon in the solar system with a large atmosphere, so it is unique." There were, of course, other instruments aboard, aside from the cameras, that would, perhaps, be more useful with Titan.

Five days earlier, or about three days before the final attitude change, the spacecraft had done the last of the mid-course corrections—the ninth since it left the Earth September 5, 1977—to refine its trajectory for the exceedingly complex series of events ahead. There had been three objectives: first, that when Voyager, going below Saturn's ring plane near Titan, moved behind Titan in relation to the Earth (an event called occultation), it would do so in line with the centers of both Titan and the Earth; second, that the spacecraft's exit a day later from occultation behind Saturn be at a precise instant; and third, that when the spacecraft passed back up through the plane of the Saturn system, it will do so in what was known as the Dione clear zone of the E ring, a zone in the thinner outer extension of the ring where its particles had been swept clear by the satellite Dione, whose orbit was 210,000 miles from Saturn. Though the mid-course correction had gone well, the accuracy of the trajectory was still not assured; because Titan was shrouded in clouds, its precise diameter, density, and mass had never been computed, and hence its effect on the spacecraft's trajectory was unknown. On that accuracy depended all the events of the next couple of days: the encounter with Titan that evening at 9:40; the close approach with Saturn the next afternoon at 3:45; and the encounters with Tethys, Mimas, Enceladus, Dione, Rhea, Hyperion, and Iapetus over the next eighteen hours.

All in all, it was a formidable series of events, tumbling on top of each other over a two-day period. And along with it—providing a sense of onward-rushing movement to the series of still photographs—was the increase in resolution, for every time the distance to an object was cut in half, the resolution doubled. Of course, the nearer the spacecraft was to encounter, the more rapidly this would happen: from Earth, which Voyager 1 had left just over three years before, it had taken a little under a year and a half for its resolution of Saturn to double just once; in the last month, resolution had doubled in two weeks and then again in one week; in the last week it had been doubling in a matter of days; and in the next couple of days it would be doubling in a matter of hours. As Soderblom put it later that day, "If you compare our resolution of Saturn's moons with our resolving power of our own Moon over the years, up until perhaps a month ago Voyager One was in the Stone Age. Then, perhaps on November ninth, it emerged into the early seventeenth century—resolution was perhaps that of the first telescopes trained on our Moon. On November twelfth, tomorrow, we'll be in the nineteenth century—we'll be able to see objects that are twenty kilometers in diameter. The day after, we'll see something akin to the Orbiter series that circled the Moon in the mid-nineteen-sixties." Then, of course, after the briefest moment, just as fast—faster, in fact, for the spacecraft will have been accelerated by the pull of Saturn—the resolution would go the other way, decreasing; and Saturn, its moons, and its rings would begin to recede, the resolution halving faster at first, then slower and slower.

Right now, Voyager 1, a day and half from close approach to Saturn, was still not seeing the moons clearly—they were all in the Antoniadi-Schiaparelli condition. The big news at the press conference was that on Tethys (which, the next afternoon at 2:16, the spacecraft would pass a quarter of a million miles from—the only encounter, and that a remote one, to occur between the close approaches to Titan that evening and Saturn the next afternoon) there appeared to be a large bright area that might, or might not, be a hill. The Saturnian moons, at this stage at least, were taking on the attributes of certain types of blocks for children—they each bore one large, readily identifiable, and identifying, image: a rosette, in the case of Rhea; a wispy spiral, in the case of Dione; and now, possibly a huge mound, in the case of Tethys. There was one other odd bit of news: the F ring, and also a small ring in one of the gaps, proved to be elliptical.

After the press conference, I visited Tobias Owen, the astronomer from Stony Brook who was head of the Titan group. "I remember, when I was nine years old, seeing Saturn with my naked eye," he said. "I was living in Santa Fe. The air was cool. I saw it in Gemini. Then, six years ago, when I was at the McDonald Observatory, in Fort Davis, Texas, standing on the catwalk outside the dome, I saw it in Gemini again—it had completed its thirty-year orbit." (Saturn was the slowest-moving planet, with the longest orbital period, of any known to the ancients; hence—according to one interpretation—the Greeks' relating of Saturn to Cronus, the Titan who is associated by some classicists with time.) In the interval vetween his two sightings of Saturn in the same part of the sky, Owen told me, he had graduated from the University of Chicago, received his doctorate in astronomy from the University of Arizona (where he studied under Kuiper), and spent a year as an associate professor of planetary sciences at the California Institute of Technology before moving, in 1970, to his present post at Stony Brook.

I asked what, for him, were the most important aspects of the Voyager visit to Saturn.

"The Saturn system has two features that are not found to the same extent anywhere else, and they may shed light on our origins," he said. "One is the ring system: Voyager gives us a chance to look at its dynamics—to study orbiting rings of loose particles that haven't formed into planetary bodies. And the other feature that Saturn offers is Titan, which Voyager will be encountering tonight. In nineteen-forty-four, Kuiper discovered that Titan has an atmosphere. It's a highly evolved atmosphere. At the same time, it can be considered a primitive atmosphere, in that it's hydrogen-rich; that is, it contains methane [CH_4]—Kuiper discovered that, too—and traces of other hydrocarbons. Ammonia [NH_3] may also be present. Whether the Earth once had a similar atmosphere is open to question, but it may have." One question that had arisen from the beginning, Owen said, was how Titan had managed to pull from the nebula, from which the Sun, Saturn, and the other planets were formed, so large a quantity of gases, and hold on to them in an atmosphere, while other similarly sized bodies, such as Ganymede, our own Moon, and Mercury, had no appreciable atmosphere. The colder temperature there is clearly part of the answer, for on the molecular level the gases are less active. There is, incidentally, nothing arbitrary about looking for, or finding, on Saturn or its moons, such compounds as methane, ammonia, water, or even organic compounds, for the most abundant gases in the nebula were hydrogen, helium, nitrogen, and oxygen—and there were considerable amounts of carbon, too. Consequently, their simplest compounds, such as ammonia (which is three hydrogen atoms and a nitrogen atom), methane (which is four hydrogen atoms and a carbon atom), and water (which is two hydrogen atoms and one oxygen atom), would be about the first things the scientists would expect to find; also high on their list would be simple organic chemicals, which are composed principally of hydrogen, nitrogen, oxygen, and, of course, carbon.

In the early 1950s, Stanley Miller, a student working under Harold Urey at the University of Chicago—while Owen was an undergraduate there—had passed an electric current (meant to simulate lightning, which would be present early in the life of a moon or planet) through a flask containing an atmosphere of methane, ammonia, water, and hydrogen, creating simple organic chemicals—a brownish gunk. (Though scientists are not certain that this is the way life started on Earth, Urey himself once said that if it wasn't, "then God missed a good bet.") Owen is substantially cautious on the subject; still, he said that he and many other scientists feel it is very likely that organic processes, if only the earliest phases of them, have taken place on Titan. This cannot be said even of Mars. Much of the driving force behind planetary exploration has always been the hope of finding some form of life, however primitive, or failing that, of organic chemistry; the argument was, and still is, that if just some of the steps toward life could have been shown to have taken place on a planet as apparently bleak as Mars (let alone our own Moon), with the slenderest atmosphere, the smallest barometric pressure so that water will not remain liquid on its surface, which in any event is saturated with ultraviolet light from the Sun lethal to life as we know it, then the chances for life, even intelligent life, on other more clement planets, in other solar systems, of which there could be a great many, would shoot up. In 1976, the two Viking landers (Owen, like many of his Voyager colleagues, had been on the Viking imaging team) found no trace of organic compounds on Mars. But Titan, Owen said, despite the cold temperatures, was far more promising than Mars, because of its thick atmosphere.

Interest in Titan increased in 1965, when Frank Low, a physicist at the University of Arizona, determined, by infrared spectroscopy, that the temperature of Titan was 165° Kelvin (−162° Fahrenheit, −108.16° Celsius)—higher than could be accounted for solely by the meager amount of sunlight that the moon received. (Scientists generally use the Kelvin temperature scale, in which the degree unit is the same as the degree unit used in the Celsius scale; however, the Kelvin zero, instead of being the freezing point of water, is the absolute zero reached theoretically when all molecular motion ceases, and hence no temperature can be lower. There are no minus Kelvin temperatures; all temperatures above absolute zero are plus. According to the Kelvin scale, water freezes at 273.16°. The Kelvin scale is particularly appropriate for defining extremely low temperatures, such as there are in the Saturn system, for even out there some temperatures are relatively higher than others, reflecting greater heat—a concept difficult to appreciate when recorded in the prodigious negative numbers that occur with the Celsius or Fahrenheit scales.) The 165° K temperature, it was thought, might be an indication that Titan's atmosphere was somehow generating a greenhouse effect, allowing heat from the Sun to pass in through it but not back out again, much as carbon dioxide does on Venus and to some extent on Mars. (Carbon dioxide was not thought to be the cause on

Titan, however; any oxygen there would have long been locked up on the surface as water ice.) In the early 1970s, however, Robert Danielson, a physicist at Princeton, hypothesized a different theory, that the high temperature reading reported earlier came from a specific layer in the atmosphere, and his hypothesis was subsequently borne out by spectroscopy. Later readings indicate that the high temperature in fact was caused by a haze of aerosols made up of hydrocarbon particles, which apparently absorb heat from the Sun. Hydrocarbons—organic compounds of hydrogen and carbon—lack nitrogen, a necessary element in the more complex organic compounds that lead to life. The hydrocarbons were thought by some to be continually raining down onto the surface.

"After Kuiper's initial discovery, a lot of us were using our own different techniques to study Titan," Owen said. "During the nineteen-seventies, when I was using an extremely sensitive spectrometer at the McDonald Observatory, I found that Titan's methane absorption bands—bands in the spectrum of sunlight that show how strongly the methane is absorbing it, and that are an indication of the gas's presence—were different in intensity from the ones from Jupiter and Saturn, and that, along with other readings, meant that on Titan, methane couldn't account for much of what we were seeing, and therefore there had to be a lot of some other gas. The problem was complicated by the dense aerosol haze, which distorted our spectrometer readings." One thing, though, seemed clear: it was the aerosols that created the yellowish-brown haze that prevented a view of the planet's surface. There was plenty of speculation about how the aerosol particles had formed. At the very least, photodissociation of methane would have occurred—that is, methane would be broken down by sunlight, its components eventually recombining to form more complex hydrocarbons such as ethane, acetylene, and ethylene. The former two were in fact detected in Titan's atmosphere before Voyager got there. Some investigators felt ammonia might be present, too; while the freezing point of ammonia, 195° K, is relatively high, they believed that at the time Saturn was forming, the temperature in the nebula around it might have been high enough for the formation of that compound, which subsequently would have become locked in the ices of Titan, and of some of the other moons as well, as what is called a hydrate (that is, united with water), where it would await release in a subsequent hypothetical warm period. There was considerable interest in this notion because if ammonia were in vapor form in the atmosphere, for which the temperature need exceed only 150° K, the photodissociation of ammonia (NH_3) into its components might provide the nitrogen necessary for prebiological organic chemistry. An ammonia atmosphere would also protect whatever complex organic aerosol polymers might have developed, by providing a shield (analogous to the ozone over the Earth) in Titan's atmosphere against the Sun's ultraviolet light, which would otherwise break them down (although eventually the shield itself would be broken down). And if the ammonia was also present in liquid form it would provide on the surface a substitute for water as a solvent, which would facilitate the creation of complex polymers, in particular by providing a nontoxic liquid environment in which their components could easily get together. However, Owen was dubious about the possibility of an ammonia ocean, or even an ammonia lake; for those, of course, the temperatures on the surface of Titan would have to be above 195° K, and they were not likely to be that high now, Owen told me, though according to some scientists they may have been in the past. The same atmospheric shield that would have blocked out the Sun's ultraviolet light would have acted as a greenhouse, holding in warmth from the Sun.

"Some of us are still hoping that Voyager will be able to see the surface of Titan," Owen said. "But it looks more and more as if the dense aerosol haze is beyond the capacity of our cameras. There have been a variety of estimates about the Titanian atmosphere—its composition, density, and pressure, which are interdependent. One of the two most popular theories is still that the atmosphere is mainly methane; this would mean a low atmospheric pressure, a small percentage of the Earth's, and thus a low density. If that turns out to be the case, there is a chance that we *could* see through it." He went on to explain that the other theory, which had been proposed by Donald M. Hunten, an atmospheric physicist at the University of Arizona, held that Titan's atmosphere was mostly nitrogen, with trace amounts of methane, and probably of other gases as well. One advantage of nitrogen is that—like carbon dioxide or ammonia—it too would tend to hold in the Sun's warmth, in a sort of greenhouse effect, thus helping to explain the high temperature readings and holding out the hope for some warmth nearer the surface. A nitrogen atmosphere would have a much higher density and pressure than one made up primarily of methane. Depending on which atmosphere, or combination of atmospheres, a scientist believed in, and how deep he thought it was, guesses about atmospheric composition were ranging from a mostly nitrogen model with a surface pressure as high as three bars—three times that of the Earth at sea level—to a mostly methane model with a pressure as low as seventeen millibars, or one-fiftieth of the Earth's. Since no one knew what the surface temperature was, either, no one could say what state (liquid or solid) the material (whatever it was) on the ground (wherever that might be) would be in.

Although the presence of hydrocarbons and, possibly, nitrogen suggested that at one time Titan might have supported prebiological organic processes similar to those that gave rise to life on Earth, no one was seriously arguing that life might be found there, because the temperature would be too frigid. Life had been pursued in far more probable places (at least with respect to temperature), like Mars, without success, Owen observed, so there was little likelihood of finding it in what amounted to a deep freeze.

At 1:30 that afternoon, in a large meeting room down a hall from the imaging team's quarters, Bradford Smith held a series of meetings of the main groups the imaging team was divided into—the Saturn atmosphere group, the Titan group, the satellite group, and the ring group. He started with the Saturn atmosphere group. "Larry Soderblom and I have been dashing around so much that we have lost touch with the science," Smith began. "We want to find out what are some of the problems, and what are some of their solutions." The group leader, Andrew P. Ingersoll, a tall scientist in his early forties, who had a high forehead, black hair, and black beard, and who was a physicist at the California Institute of Technology, said, "We'd like to confirm the width of the horizontal, colored bands on Saturn." An older scientist with a pink face and whitish hair, Verner E. Suomi, said, "Another issue is over the reality of the speed of the winds at the equator, the equatorial jet, which is quite wide, and which we think is moving at least four hundred or five hundred kilometers an hour. It would be interesting if it was roaring along that fast. The question of dynamics is interesting: How do you get the energy there? The winds on Jupiter are not as fast. Another problem is that as we get away from Saturn's equator, it's even harder to judge the speeds of the winds because of the increasing curvature of the planet—it's very frustrating." Wherever they looked, the problem remained the one of finding some sort of structure within the clouds definitive enough and permanent enough to be tracked. Ingersoll, taking up the problem of Saturn's curvature, pointed to some smaller bands about fifty-five degrees north and south of the equatorial region, on a recent picture of the planet, and said, "I wonder how you get energy into these little belt zones? The fine structure inside looks like chevrons."

Reta Beebe said that—despite all the problems—it might indeed be possible to measure the wind speeds in those high-latitude bands by tracking the movement of the chevrons in two photographs taken at different times. Garry E. Hunt, another atmospheric physicist—a large Englishman from University College in London—nodded and said it might be useful to make some comparisons with wind speeds on Jupiter. It might be particularly helpful for the press. Another team member present shook his head and said that the Public Affairs Office was against comparing the two planets, because in most comparisons Saturn would look bad: Jupiter, and Jupiter's moons, were so much bigger. "But Saturn's winds are faster," Hunt said. "Do you want me to do a Saturn-Jupiter comparison of the winds?" Smith said he didn't know; the purpose of the meeting was to find out what was going on, not to make decisions.

Next it was the turn of the Titan group. As Titan had such a thick atmosphere, in some respects even more intriguing than Saturn's, most of the Saturn atmosphere people stayed. "Tell us what is going on," Smith said. Though Owen, the group leader, was there, he liked to stay in the background and let others do the talking. James Pollack, Cuzzi's associate from Ames, a slender man in his early forties, said, "We're still trying to see what Titan has to offer us. Things have been disappointing as far as seeing the structural details in the atmosphere that we would like to see—altitude differences, dips and troughs, that sort of thing. It's too early to throw in the towel on those features, though; the methane filter will be our best chance."

"We have lost the methane filter," someone in the room informed him. "In the sequence already loaded into the spacecraft computer, they've put in another filter instead."

"That's incredibly unlucky," Pollack said. (As it happened, the omission made no difference, for on a later series of pictures with the filter, the clouds were uniform and featureless.) Pollack went on, "To me, the most exciting thing is the thin layer of haze, which now seems to be visible high over Titan, outside the yellowish aerosol layer." This had appeared on enhancements of more recent photographs of Titan's limb, the outer edge of its disc, which gave the best cross section of the atmosphere, as it was viewed from the side—and the higher up, which is to say the farther out, the clearer, for there was increasingly less atmosphere. Pollack suggested the high haze layer, visible as a faint aureole around the satellite's outer rim, might be a condensation phenomenon such as was seen in the Martian atmosphere, at a high altitude where the temperature was very cold. "As we get closer, we may see some structure inside this haze layer," he said.

Smith wanted to know why the northern hemisphere of Titan was darker than the southern hemisphere. Verner Suomi suggested that the haze layer in the north, where it was now spring, might be thinner, allowing the brownish-yellow of the aerosol layer farther down to show through more. Alternately, he said, perhaps the atmosphere in the north was rising and the atmosphere in the south was falling, which would bring the aerosols higher in the atmosphere in the northern hemisphere and push them down in the south. (In the eighteenth century, a similar system, known as a Hadley cell, had been proposed for the Earth's atmosphere—a term the Voyager scientists revived now.) Both these theories—along with Owen's of the day before, explaining the darkening by Saturn's magnetic field hitting Titan differently from the north than from the south—had a plausible ring; clearly, with respect to Titan, the scientists were very much in the third of Cuzzi's four stages of confronting a scientific unknown—thinking of new physics or new theories that might apply—and they seemed to be enjoying it.

Before the satellite group session started, there was a general exodus of atmospheric scientists—the Saturnians and the Titanians. Smith asked those who were left, "Anybody mind if there are TV cameras in here?" Soderblom, the head of the satellite group, said, "Sure, why not? We can't tell the difference anymore." He started to do up his tie and to comb his hair, looked startled when he heard a couple of snickers, then undid his tie and rumpled his hair. "That'll make them think this really is a working session," he said. With the camera in place in a back row, Merton Davies, one of the original imaging-team members, a tall, slender man who looked like Gary Cooper and spoke in the same slow drawl, reported on some preliminary measurements he had made of some of the satellites' shapes, based on their reflectivity. As an expert in geodesy, the measurement of a body's size and form, Davies needed as many pictures of a satellite, from as many angles, as he could get; and consequently, in the year or two before the encounter, when the scientists had been deciding what pictures to take—an interminable process known as "sequencing" or "targeting"—Davies, whose needs were perhaps greatest and most readily definable, had been one of the toughest scrappers.

"I have some measurements on Rhea and Titan," he began. "Mimas doesn't look like it's quite round."

Someone asked, "How about Enceladus?" This moon was of particular interest because of a theory offered by Charles Yoder, an expert in celestial mechanics at JPL. Yoder proposed that for various reasons having to do with the eccentricity of its orbit and the resulting effect upon it of Saturn's gravitational, or tidal, pull, the satellite might be warm. The warmer it was, the more spherical it would tend to be, as it would be more plastic and therefore would smooth out or adjust away any irregularities. The fact that Mimas wasn't quite round suggested that it was more rigid, and hence very likely colder.

"I don't know about Enceladus's shape," Davies replied. "I haven't been able to get much on Enceladus." That satellite was not a prime target of Voyager 1. It had been deemed of such great interest that Voyager 2 was scheduled to pass much closer to it—and consequently, in targeting Voyager 1, Enceladus had been sacrificed frequently to other objectives. The loss of this imagery of Enceladus dismayed Davies, who had argued that Voyager 1 would not only be seeing the satellite from different angles from Voyager 2, but also should make a point of imaging the satellite anyway, for redundancy, in case there was some mishap with Voyager 2. In addition, the more pictures of Enceladus, the more precisely Davies could do his geodetic work.

Davies went on, "Here's one of Hyperion; it's so far away you can't see much, but it looks as if it might be a little off-round, too."

Soderblom began making a list on the blackboard of all the moons, and how close to the spacecraft they would be when it photographed them in the next couple of days. "We have a real chance of missing some of them," Smith mused out loud; he was, he said, increasingly concerned that the unknown mass of Titan would throw off the cameras for all subsequent encounters.

As the members of the ring group were so awash in new photographs that none of them could be pried away to make a report, Smith adjourned the meeting. I walked out with David Morrison, a dark-haired, chatty scientist from the University of Hawaii, who said, "I'm an astronomer going through the experience of seeing these moons change from astronomical bodies to geological bodies. Right now, I know as much about them as anyone. Before long, I'll be obsolete—the geologists will know more. The moon I'll be concentrating on most probably will be Iapetus, the outermost next to Phoebe, which has intrigued astronomers for centuries because, of all the moons in the solar system, it has the greatest changes in reflectivity: sometimes it is very bright, sometimes it is so dark you can hardly see it at all. Also, the closest Voyager One will get to it is a little over one and a half million miles, so the chances are good that it will remain an astronomical object a little longer—at least until August, when Voyager Two will come within a third of that distance. We got our first good photo of it just four days ago; the reason for the change in brightness turns out to be that one side is dark and one side is light, and so it depends what side you're looking at. It's obvious that one side is water ice, while the other side is rocky or dirty material." From a folder he was carrying he produced a hard copy. Soderblom, a geologist, looked over his shoulder. He appeared interested, but—for the time being at least—he had no comment to make.

In one of the interactive rooms, two graduate students in planetary science (a relatively new field that borrows from astronomy, physics, and geology) at the University of Arizona were stretching a new picture of Tethys. Nicholas M. Schneider and John R. Spencer had recently arrived with their professor, Robert Strom, a planetary geologist on the imaging team. "We convinced Strom that our presence here would be absolutely essential," Spencer said, as an enhanced view of Tethys popped onto the screen. Essential or not, they were clearly having the time of their lives; in particular, they liked to run the interactive machines—a job the older scientists were glad to let them do. Schneider, a very tall, very thin young man with a leonine head of hair, said, "See the circle—that they thought yesterday was a hill? It's not a hill anymore—we're now calling it simply a 'circular feature.' It's probably tectonic; that is, of internal origin."

"We use those terms when we don't know what else to say," said Spencer, a wiry, dark-haired young Englishman, a little shorter than Schneider. "Even so, that we can apply a word like 'tectonic' to these moons at all is a step forward—it means we geologists are beginning to come into our own." Unlike Morrison, who had sounded wistful at this turn of events, Spencer was clearly pleased. One rarely heard a geologist credit astronomers for their centuries of studying the planets, which had paved the geologists' way. Tethys, though, had by no means yet passed

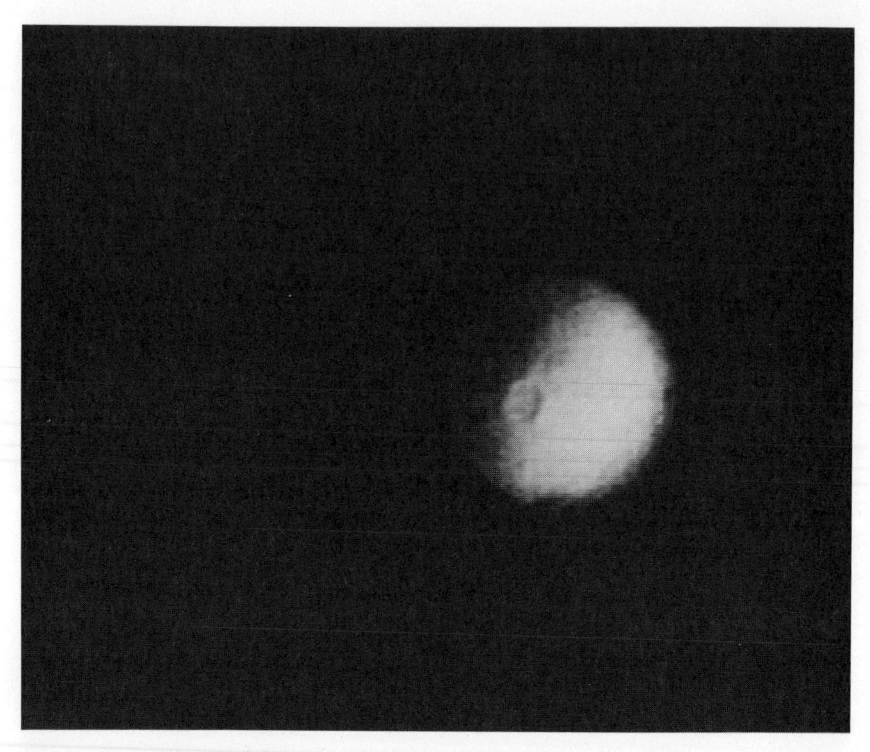

The "circular feature" on Tethys. Imaged November 11, 1980.

out of the astronomers' domain, for it was still sufficiently in the Schiaparelli stage to have developed a mysterious thick black band, apparently running north and south between its poles, about which the geologists were remarkably silent.

Although most imaging-team discoveries at this stage were being credited to the team as a whole, and not to any individual, the ringlets that I had heard were eccentric—all the rest are virtually perfect circles—had in fact been identified by an astronomer at JPL named Richard Terrile. At twenty-nine, Terrile was one of the youngest members of the imaging team, and his work was an example of what Cuzzi had meant when he talked of younger scientists' being freer to pursue lines of inquiry that caught their eye. Terrile, a former student of Owen's at Stony Brook, is a tall, thin, swarthy man with dark-brown hair and a dark-brown mustache.

In his office, across the corridor from the interactive rooms, I asked him how he had discovered the eccentric rings.

"Making discoveries is just a matter of who's around," Terrile began modestly. I had been told that Terrile was one of the hardest-working members of the team, and consequently was around a great deal. "A few days ago, Torrence Johnson—another member of the imaging team—and I were here when a picture of the F ring came down," Terrile continued. "Because the orbit of its outer shepherding moon, S-Thirteen, was so eccentric, we thought the F ring itself might be eccentric, too, and when we compared two pictures of it showing its relationship to the outer edge of the A ring, we saw that it was. Then, last night, I saw an eccentric ring in one of the small gaps in the C ring. I noticed it because it was easy to see the ring on one side of Saturn, where it was near the center of the gap, and hard to see on the other, where it had

One of Terrile's discoveries: the eccentric ring in the gap within the C ring, whose wobble is demonstrated when two pictures, from opposite sides of the ring, are fitted together, the black line being where they meet; the features match up, with one notable exception. From imagery made November 10, 1980.

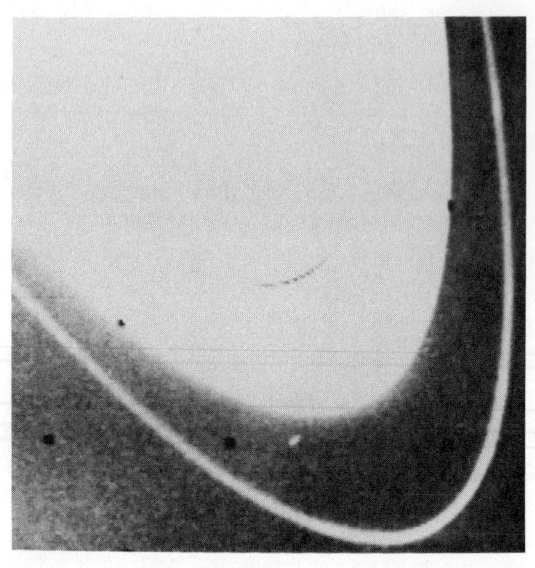

S15, the A-ring shepherd: another of Terrile's discoveries. Imaged November 6, 1980.

moved over to one side, closer to some other ringlets. We now had two eccentric rings. An hour or so later, I decided to look inside the Cassini division—and, yes, there was an eccentric ring *there*. So it seems that eccentric rings and gaps go together. Kind of fun!"

A month earlier, Terrile had been the first scientist to notice the B-ring spokes. He had also been the first to spot S15, the moonlet guarding the outer edge of the A ring. These discoveries, one of his colleagues suggested, resulted from a dual ability to ask the right questions and to use a sharp pair of eyes. "It's a fantastic time to be in the field!" Terrile told me. "Last night, I had dinner with some friends. They wanted to go out for after-dinner drinks. I said I'd join them later; I had to go to the lab. They said, 'Don't come back until you've found something new'—and, sure enough, I was able to oblige them."

I asked Terrile his own opinion of why he, perhaps more than anyone else on the team, had made so many discoveries. He shrugged. "I don't know," he said. "Maybe I've got an advantage because, as an astronomer, I'm an observer. I did my thesis with the two-hundred-inch telescope at Mount Palomar. In astronomy, you're looking for very subtle, faint things—your brain suppresses noise."

I asked him how he had gotten into astronomy. "I grew up in Flushing, in Queens. During my whole childhood, I was interested in science.

I had a paste-up science scrapbook. I watched 'Mr. Wizard,' who used to do physics experiments on television. When I went to the John Bowne High School in Flushing, I took special programs in physics and biology. I went on to take a double major at the State University of New York at Stony Brook—physics and astronomy. Deciding for astronomy, I went on to the California Institute of Technology in planetary science. Cal Tech is the place to go for that; the best facilities on *this* planet to study other planets are available there: Mount Palomar and, of course, JPL. I prefer planets to stars—they are things you can touch, maybe even walk on. It's hard to believe I'm being paid for the fun I'm having right now. It's like watching a science-fiction movie—no, it's like being *in* a science-fiction movie, only it's not a movie, it's real: we are going into new systems and discovering new worlds."

The closest approach to Titan would take place at 9:49 that evening, and the crossing of Saturn's ring plane would occur at 11:22. Despite the full agenda, when I went back to the imaging-team quarters after dinner I found hardly anyone there. Tobias Owen was one of a small group watching the raws come in. "Oh, gee, I hope Titan shows up better than that later," a technician said. The raw on the screen was blurry, and since Titan itself was something of a blur to begin with, the result was not very informative. Owen said he thought it was now definite that, because of the thickness of Titan's haze, they wouldn't be able to see down to the surface. The encounter was only an hour off, but it would take almost an hour and a half after that for the raws of the encounter to arrive at JPL, and two hours after that for the pictures to be put in the disc packs so they could be enhanced on the interactives. Owen looked at his watch, yawned, and said he thought he would go home soon and try to come in early the next morning. Exploration by unmanned spacecraft is a delayed-action affair. As the spacecraft sped toward what is doubtless the most interesting moon in the solar system, the members of the imaging team, one by one, went home.

While many scientists went home to sleep, the night shift of flight controllers, sequestered inside a glass-and-concrete building about fifty yards uphill, of course did not. I found out later from Richard Laeser, the mission director—a big, broad-shouldered man who smiles frequently when things go well—that the night of the encounter with Titan was one of close calls. For one, the thunderstorms that had been hovering for days over the radio dish in Spain cleared up only just in time to enable it to receive the encounter data. For another, although the flight controllers' estimate of the unknown mass of Titan, as expected, had been a little off, luckily it was not far enough off to throw the spacecraft seriously onto the wrong course. "As it happens, the big surprise that night came from a different quarter," Laeser said. "Titan itself turned out not to be where we thought it was—its center point was a hundred kilometers away. If we had known that at the time of the final mid-course correction last week, we would have aimed differently. However, it turned out that when we cranked those miscalculations in on top of the one about Titan's mass, they to an extent balanced each other out, for we missed the spot we were aiming for, the one behind Titan in a direct line with the centers of both the satellite and the Earth, by only twenty kilometers. We could have missed by more and still been on target for the rest of the encounter, because we had two hundred and sixty-five miles' leeway." Still, the targeting information for aiming the cameras during the encounters of the next couple of days—previously stored in the spacecraft computer—had to be corrected, which the flight controllers could do by sending up to the computer what they called an "overlay" for the data already there. "We had known right along that Titan would fool us, particularly as we would be flying almost at the cloud tops, and accordingly we had held a number of rehearsals to see how quickly we could figure the changes and shoot them up to the spacecraft," Laeser said. On the night of the encounter, they figured the corrections for each target, made the overlay, and sent it up within two hours after passing Titan—twelve hours before the new information was needed for operations and photography close to the planet. With the overlay in and Titan safely passed, Laeser now had good reason to think that the spacecraft would meet accurately the next two major milestones on its trajectory: the exit from behind Saturn and the recrossing of the ring plane.

The next morning—Wednesday, November 12th, the day of the encounter with the planet itself—the bulletin board in the central mall of JPL said that today the spacecraft would come to within 77,202 miles of Saturn; that its velocity, close to double what it was the day before, would by then be 56,559 miles per hour; and that the days to go were zero. A

A replica of the Voyager spacecraft in the Von Karman Auditorium at the Jet Propulsion Laboratory, looking as it might in space. The boom with most of the science instruments is on the right; the two cameras, along with the ultraviolet and infrared spectrometers and the photopolarimeter, are on the steerable platform hanging from the end. At the end of the long boom on the left—a third of the way around the spacecraft—are the magnetometers; and on the little boom below that—in reality a third of the way around from the other two booms—are the three radioisotopic thermoelectric generators. The white radio dish at the top would normally be pointed at the Earth and hence the Sun, so that the ten-sided box underneath, which contains the craft's fuel and computers, and from which the three booms jut, is usually in the shade.

Photo of Saturn—somewhat more luminous than its matching ocher rings—taken by Voyager 2 from thirteen million miles away, as it approached. Tethys, Dione, and Rhea (left to right) are orbiting at the bottom, white specks against the black of space, while Mimas is barely visible against the planet, above and to the left of Tethys. The big black spot on the planet is Tethys's shadow, and the blackish smudges on the rings are the spokes.

The rings of Saturn, from outside in: the thin strand of the F ring; the A ring, broken near its outer edge by its big gap; the Cassini division; the thick, broad B ring; and the thin, ripply C ring. Saturn—nine hundred and thirty thousand miles from Voyager 1 when this scene was imaged on November 13, 1980—is visible through the rings, which have cast their shadow on the planet.

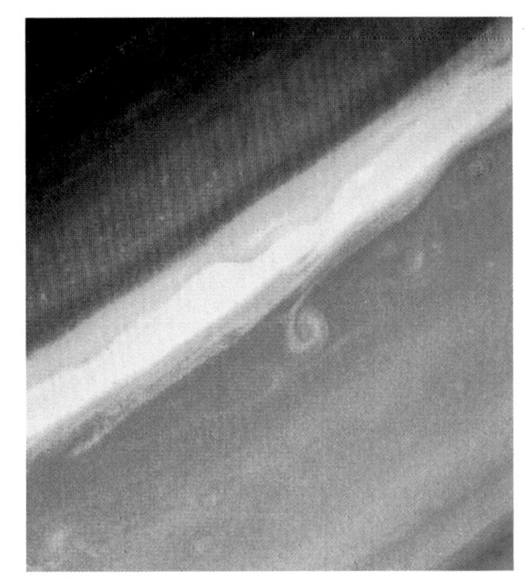

The clouds of Saturn, as seen through the high-resolution camera of Voyager 2, which brought out more detail than Voyager 1's cameras. The mysterious 6, presumably a wisp from the ribbonlike white feature above and rolled into its spiral shape by the shear of opposing wind currents, is at the center. Brown spots and white spots dot the landscape, or the cloudscape; as the picture was taken from 5.8 million miles away, the smallest feature visible is about fifty-six miles across.

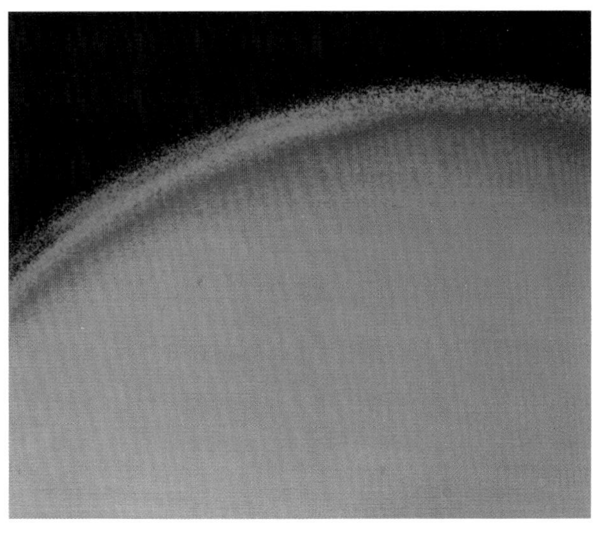

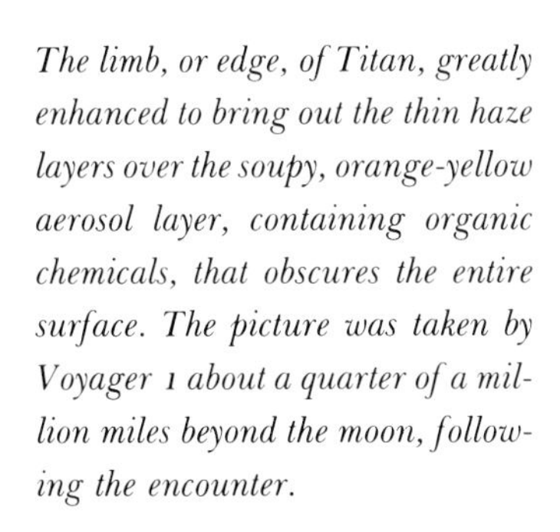

The limb, or edge, of Titan, greatly enhanced to bring out the thin haze layers over the soupy, orange-yellow aerosol layer, containing organic chemicals, that obscures the entire surface. The picture was taken by Voyager 1 about a quarter of a million miles beyond the moon, following the encounter.

Photograph of Titan, eclipsing the Sun, taken by Voyager 2 from a little over half a million miles away. The manner in which the sunlight is scattered all around the limb indicated to Owen that even the high, thin haze layers contain particles.

Mimas: 116,962 miles from Saturn, 242 miles in diameter. Although it is one of the smallest of the previously known moons of Saturn, its gravitational effect on the rings is the greatest of those moons, because it is the closest of them. The giant crater is on the far side. Voyager 1 imagery taken November 12, 1980, from a distance of eighty thousand miles.

Enceladus: 147,918 miles from Saturn, about 310 miles in diameter. Its variegated appearance—more so than any other moon in the solar system, with the exception of Jupiter's Io—indicates that it has been warmer than any of Saturn's other moons. Voyager 2 imagery, taken August 25, 1981, from a distance of 74,000 miles.

Tethys: 183,108 miles from Saturn, about 659 miles in diameter. Its giant crater is out of sight on the far side, but the giant trench is curling down to the right. Voyager 2 imagery taken August 25, 1981, from a distance of 368,000 miles.

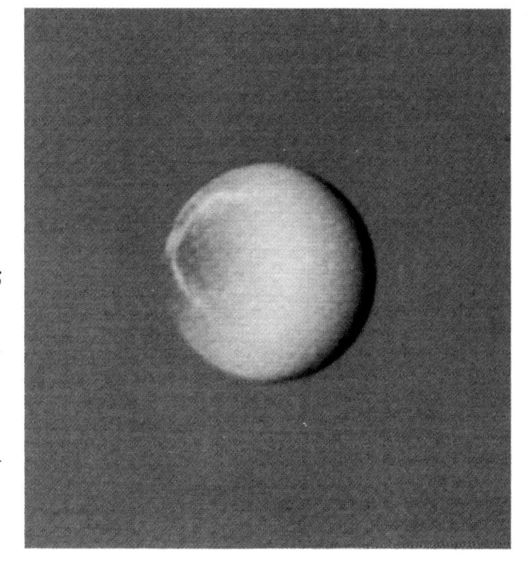

Dione: 234,528 miles from Saturn, about 696 miles in diameter. The moon, photographed passing in front of Saturn, is similar in size to Tethys but shows more signs of having once been warm. The wispy terrain, very likely one of those signs, is to the left, on the trailing hemisphere. Voyager 1 imagery taken November 11, 1980, from a distance of 234,000 miles.

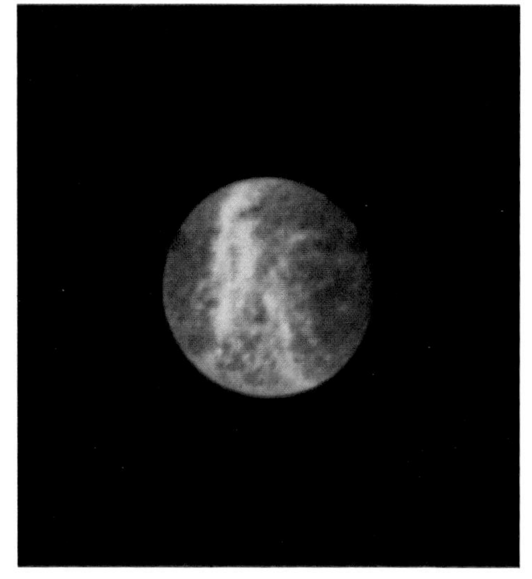

Rhea: 327,540 miles from Saturn, about nine hundred and fifty miles in diameter. The color is somewhat exaggerated to bring out the wispy terrain, which Rhea shares with Dione. Very likely, on both moons, it is a welter of fractures whose edges have been whitened by outgassing from within the body. Voyager 1 imagery taken November 11, 1980, from a distance of one million miles.

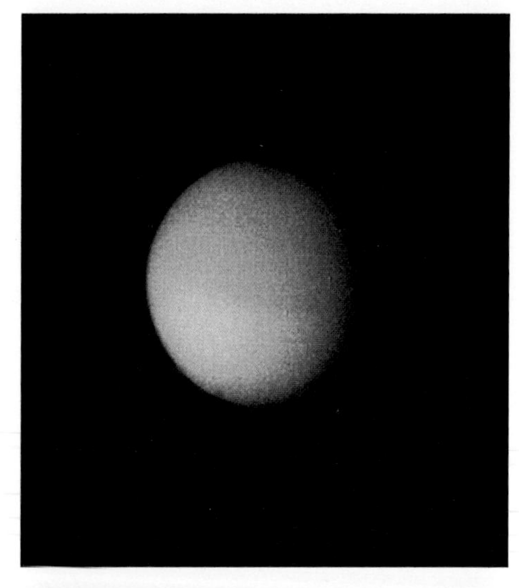

Titan: 759,263 miles from Saturn, a little over 3,196 miles in diameter. Titan is by far Saturn's biggest moon, one of the biggest in the solar system, and the only one with an appreciable atmosphere; thicker than the Earth's, it is composed in part of aerosol particles. From this distance, about 1,560,000 miles, the darkness of the northern hemisphere is apparent, as is the dividing line, possibly paralleling the equator, between the northern and the lighter southern hemisphere. Voyager 1 imagery, taken November 11, 1980.

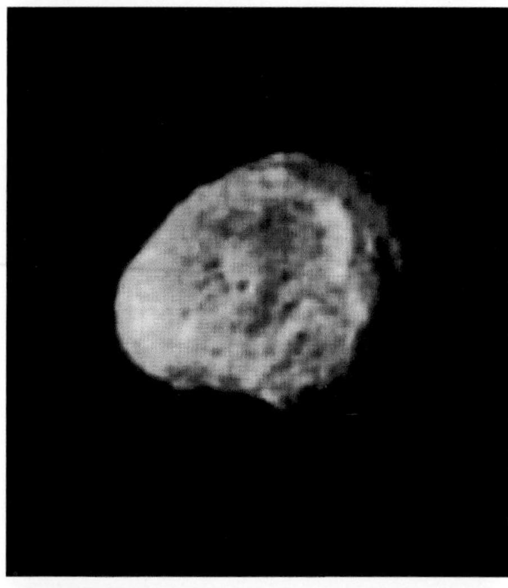

Hyperion: 920,293 miles from Saturn; its dimensions are approximately 255 miles by 162 miles by 137 miles. Its irregular shape is probably the result of repeated impacts. Voyager 2 photograph, taken August 24, 1981, from a distance of about three hundred thousand miles.

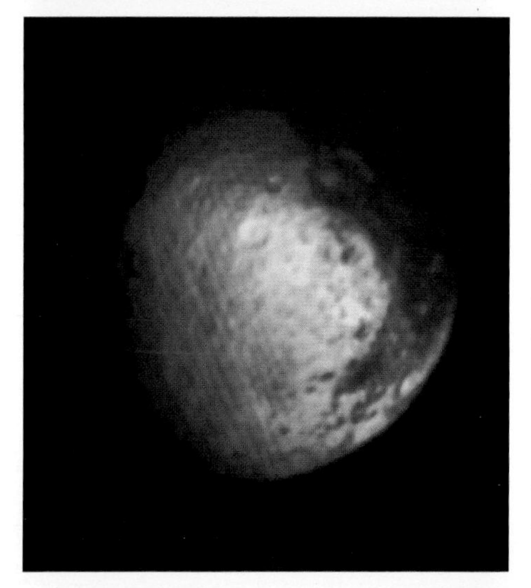

Iapetus: 2,212,681 miles from Saturn, it is the most distant of the planet's regular satellites; Phoebe, 8,049,615 miles away but in an orbit highly inclined to those of the other moons, is thought to be a captured body. With a diameter of about nine hundred and seven miles, Iapetus is the third largest of Saturn's moons, after Titan and Rhea. With its black carbonaceous markings and its icy whites, Iapetus has the greatest contrast of any body in the solar system. Voyager 2 imagery taken August 22, 1981, from a distance of 680,000 miles.

This artificially colored view of Saturn's rings, imaged by Voyager 2 on August 17, 1981, but not processed for several days, indicates that the ice particles in different regions contain traces of different chemicals, suggesting to one Voyager scientist, Eugene Shoemaker, that the three major rings might have been caused by the fragmentation of three parent bodies. Other Voyager scientists believe the rings formed as a residue of the nebula from which Saturn itself coalesced.

Farewell to Saturn: One of the first pictures taken by Voyager 2 after its damaged scan platform, with the cameras, reacquired Saturn on August 29, from 2.1 million miles beyond the planet. The spacecraft is now below the ring plane, giving a good view of Saturn's southern hemisphere.

small green-and-white pavilion had been set up at the entrance to JPL to accommodate visitors—a number of dignitaries were expected. Upstairs in the imaging-team quarters, all was chaos. The resolution of the Saturn system was increasing dramatically. In one of the interactive rooms, David Morrison pulled a hard copy off the machine. "Have you seen the new picture of the F ring?" he asked, holding it out to me. Before someone else snatched it away, I had just time to see that the ring appeared to break here and there into two separate strands that wound around each other. These were evidently the clumps that had been noticed earlier.

To everyone's amazement, the F ring appeared to be braided. Richard Terrile, who was taping up a mosaic of it, muttered, "Pretty amazing!" In his mosaic, which showed a longer stretch of the ring than Morrison's picture, it looked interwoven much of the way around.

"If the music of the spheres comes from Saturn, here's a frayed string," someone said. (It was a metaphor upon which endless changes could be rung, and indeed were; another, contrived by one of the more esoteric scientists, had to do with the fact that composers have already written music based on resonances in the solar system—and perhaps would do the same now with Saturn.)

In the other interactive room, scientists were looking at a picture of the rings taken a couple of hours before—from underneath, for the spacecraft was now below the ring plane. The patterns of brightness and darkness were somewhat different from what they had been in pictures taken above the ring plane, where light from the Sun, which was a few degrees above the rings, was reflected from their surfaces. Extremely opaque rings, such as those in the center of the B ring, had glinted in those pictures but were dark now, while thinner regions, such as the C ring, had been comparatively dim when seen from above but were noticeably brighter than the opaque rings when seen from below, because of the scattering of light. Truly vacant areas were dark when seen both from above and from below. Cuzzi, who was one of the observers, told me he thought much could be learned about the relative optical thickness of a ring—which is related to particle density—by comparing the photographs from the two sides, ring by ring and gap by gap.

Out in the corridor, Harold Masursky was holding a picture and saying, "Incredible, incredible."

I asked him what he had found.

"We had a violent argument yesterday over whether some vague markings on Tethys were craters," he said. "Some scientists argued they were patterns—light reflections. I was one of the ones who held out for

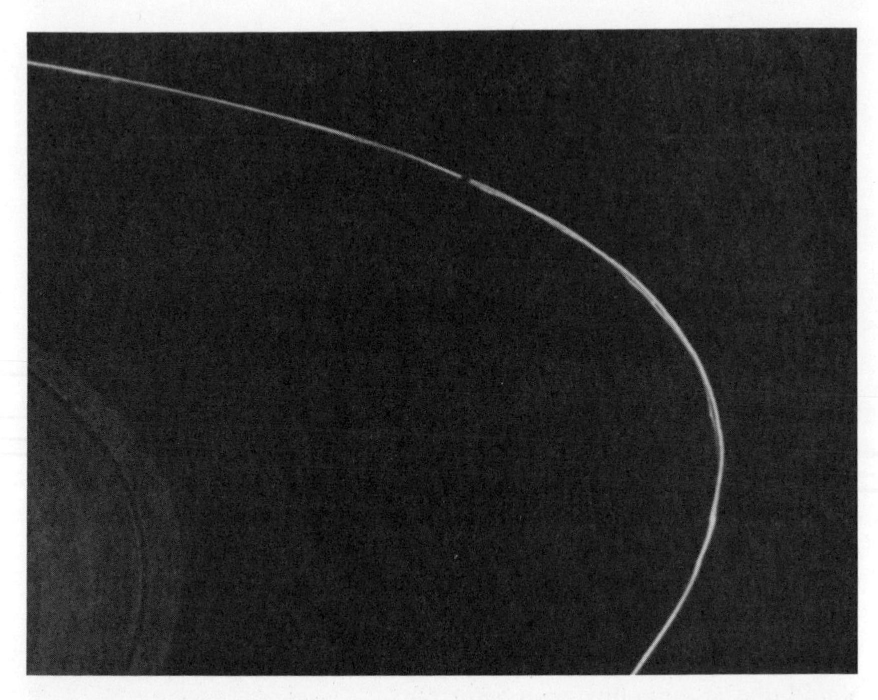

ABOVE: *Frayed string: the braided F ring, imaged November 11, 1980.* BELOW: *This picture is an enlarged segment.*

craters. And now look!" He handed me a hard copy of a picture of Rhea showing several large craters. "I don't know what should be so surprising about finding craters on these moons, but some people have to be knocked on the head. And Rhea seems to have some ropy bundles and wispy terrain, like Dione's—the same family of things we saw on some of Jupiter's moons. Hang on, though. We'll get pictures of Rhea between a hundred and a thousand times as good as these. We'll come so close that we'll do a seven-picture high-resolution mosaic of Rhea—it will take that many frames to get it all in! That will be something worth waiting for." Voyager 1 would come closer to Rhea than to any other body in the Saturnian system except Titan; during the night it would pass Rhea just forty-five thousand miles from its apparently crater-studded surface. "Now look here on Mimas," Masursky went on, handing me another hard copy. "Mimas this morning had a clear crater." The crater on Mimas was indeed clear, and impressively large. It was, he told me, eighty miles across, a third of Mimas's diameter; a huge, bullet-shaped peak, apparently as high as the crater's rim, rose from the floor. This peak, Masursky said, had been caused by a rebound of the solid ice, which under a meteoric impact would bounce up with some elasticity; in this respect, ice at those low temperatures would behave like rock. Another gigantic structure had just turned up as well—a vast trench on Tethys, running north and south and at least six hundred miles long, possibly longer.

I asked Masursky, who for the last fifteen or so years has made a specialty of mapping the Moon, Mars, Mercury, and other planets or moons, whether—with so many curious features showing up on the surface of Saturn's other satellites—he was disappointed that there had been no rifts in Titan's clouds, allowing a glimpse of the ground. "No," he said. "You have to take the bad news with the good. And I never really thought Titan would be visible. We didn't even have money budgeted to make maps of Titan! But there is a silver lining to those clouds: they may help boost that Venus mission we're hoping to get funded, which would carry among other things radar for mapping Venus's cloud-covered surface; if the device would be useful also for a Titan mission, that would be an additional reason for developing it." NASA scientists and executives, even at the height of one mission, were always worrying about the next—indeed, they often tried to parlay a present success into a future mission—something they were doing with greater urgency today, now that funding for future programs was in so much doubt.

There was such a babble of voices in the corridor, outside the interactive rooms, that Brad Smith's voice boomed over a loudspeaker, "Will everyone please cool it?" Masursky retired to his own office.

Mimas's giant crater. Imaged November 12, 1980.

The vast trench on Tethys. Imaged on November 12, 1980.

"I was kind of upset yesterday, because sometimes we couldn't agree on what we were seeing," Masursky continued. "There are some features on the moons that we really *can't* identify. But we've already reached the point where we have to begin naming features, so that we can talk about them."

Names for new celestial objects and for the features on them have to be approved by the International Astronomical Union, which has created nomenclature committees for that purpose. Scientists who discover new satellites submit names for consideration, but within categories prescribed by these committees. The solar system has been parceled out among the committees; Masursky is on the committee for the outer solar system, which is chaired by Tobias Owen. The IAU has a very rigorous, conservative set of guidelines; they stress the importance of dignity and universality and ban the use of names of living people.

I asked Masursky how the committee went about picking names.

"We try to be consistent—to pick related names for the moons in a system or for the features on one moon," he said. "We often use names from myths. For Callisto, the outermost moon of Jupiter, and an obviously icy body, we used names from Scandinavian sagas. As we moved inward toward Jupiter, we moved from myths of northern Europe to myths of southern Europe. We made an exception with the volcanic moon Io; there we went to volcano and hearth gods, regardless of where they came from. As for Saturn, we don't know what we're going to do yet."

The nomenclature meetings, I had heard, could be stormy. "We're in something of a box, because in the past we have been saying you can't put names on features unless they are solid topography that we can identify—yet now we are under pressure to get on with it, just so we can talk to each other about what we're seeing," Masursky said. "The danger is, though, that if you put the wrong name on a feature, which can happen if you don't know what it is but you go ahead and name it anyway, not only might the name be misleading, but, worse, you're stuck with it, because people are very conservative about names. There are still people who refer to the big volcanoes on Mars—Olympus Mons, Ascraeus Mons—as North Spot or South Spot, which is what we called them when they first appeared as black marks through the Martian dust clouds. One suggestion has been, before we commit ourselves to a name, to designate objects by what we call alphanumerics, a nondescriptive combination of a letter and a number, such as we have used for Saturn's new satellites, S10 through S15. But alphanumerics don't have the same appeal—and besides you get stuck with *them*: people get emotionally attached to the first name they hear; they may be a kind of status symbol, because they suggest the user was present at the Creation; and years from now, long after they have all received proper names, there will be those who will refer to the co-orbital satellites, say, as S10 and S11. So you really can't correct a mistake in nomenclature. Nomenclature ought to be an accurate index of what we know. If you get squashy on nomenclature, and vague, it means you're uncertain—and it means you don't have a good handle."

Standing in the back of one of the interactive rooms, Andy Ingersoll, the tall, bearded, Saturn-atmosphere man, was peering at the interactive, trying to see the screen between and around the heads of about a dozen colleagues in front of him; he was having only marginal success. I asked him if there was anything new with Saturn itself. "No, not yet," he said. "It's still too early. The problem is that Jupiter spoiled us! It is so much larger, and the features on it so much bigger and so much clearer, that we had two months before the encounter when we had wonderful imagery, with things on the planet's surface we could actually track and follow for long periods. Here, it's only been in the last week that we have been able to see any features at all, and that hasn't given us the time to measure the velocities of the winds. Now we don't have the leisure, though we are beginning to notice some things that puzzle us: On Jupiter, with respect to the latitudes north and south of the equator, there was a correlation between the light and dark bands and the directions of the winds in them. That doesn't hold true for Saturn, it seems: the bands and the winds didn't seem correlated. We're wondering whether that would have to do with the fact that Jupiter has no seasons—it is inclined only three degrees to the ecliptic. The seasonal differences on Saturn, which has a greater inclination even than the Earth's, are much sharper than they are here. Also, of course, the rings shadow the winter pole. All this affects the heat-input into the planet, which drives the winds and affects the wind patterns. There is, of course, a major difference among scientists about the causes of the winds and patterns: there is the meteorological view, which holds that the major driving force comes from the outside and that everything happens in the top fifty kilometers of the atmosphere; and the geophysical point of view, which holds that everything is affected by action deep down inside the planet—that the energy that drives the winds comes from there. We had these arguments over Jupiter, and now they're looming over Saturn."

Jupiter, apparently, gets twice as much energy from the Sun as Saturn does, but its winds are slower. It radiates into space somewhat more energy than it gets from the Sun, the reason being that it has its own heat source: at the time of its initial formation from the solar nebula and its subsequent collapse to its present hefty size, it generated enormous amounts of heat, and calculations show that the amount of heat remaining today, over four and a half billion years later, would be enough to account for the heat radiating from Jupiter, over and above the amount it was getting from the Sun. Saturn was radiating into space about twice as much energy as it gets from the Sun—relatively much more than Jupiter was doing—and this argued that it had an additional internal source. Since Saturn is a smaller body than Jupiter, it would have lost much more of its original heat—enough, according to a theory that Ingersoll liked, so that about halfway to its center, the helium, which after hydrogen is Saturn's second largest component, would liquify into droplets and, being heavier than the hydrogen, would fall through it, generating heat by friction. The interior of Saturn halfway down is like nothing known to us; its pressure is so great that the hydrogen the helium is falling through is also in a fluid state and, farther down, in a metallic state. (In cross section, Saturn was thought to have an outer atmospheric layer approximately a hundred miles deep; below that, the gases got progressively denser until they became fluid at a depth of a few hundred miles; about halfway down, the hydrogen became metallic, while remaining, of course, in a somewhat fluid condition; at the center was a core, about twice the radius of the Earth, of ice and silicate rock; though the temperature at the center was later estimated at 15,000° K, the pressure was so great the rock was most likely liquid.)

Despite the great heat at Saturn's core, the helium dropping through the hydrogen—or some other process—was needed to explain the temperature at the surface, which was 97° K. At Jupiter, one of Voyager's instruments, the infrared spectrometer, found that the amount of helium near the surface, in relation to hydrogen, was the same as in the Sun and, presumably, in the solar nebula—which was in accordance with the theory, for on Jupiter the helium was not supposed to have precipitated. On Saturn, though, if the theory was correct, there should be a depletion at the surface in the amount of helium in relation to hydrogen; and indeed that instrument would be getting information on Saturn's helium-to-hydrogen ratio very shortly.

Wherever the energy came from, said Ingersoll, who himself was becoming more partial to the internal theory, it would set up convection currents in the atmosphere; that is to say, from deep inside Saturn, masses of warmer material would rise and masses of colder material would sink, and the vertical updrafts would be translated into the horizontal winds observed at the surface. How this happened, nobody knew. The matter, though, was becoming more and more enticing, for the estimates of Saturn's winds at the equator, where they are the fastest, kept being revised upward since yesterday, when it had been around three hundred miles per hour; today, it was up around five hundred miles an hour.

I asked whether the disputes between scientists were speeding up, too, now that more data was arriving. Compared to earlier missions to the Moon or Mars, when there had been considerable disagreement ahead of time over what to expect, the mission to Saturn had seemed relatively calm.

Reta Beebe, who was standing next to Ingersoll, said, "Oh, no. We're getting so much data now it takes all our energy just to integrate it."

"We're getting so much data, it's hard to get away from the facts," Ingersoll said. "It's when you're far away from getting the data that the wild ideas are likely to occur."

"What frequently happens around here is that if two people start disputing, a third person will introduce some new fact that renders each side obsolete, and that's the end of it," Beebe said.

Evidently, with Saturn, the Schiaparelli-Antoniadi phase was coming to an end; Saturn with its rings and moons was being rocketed into the space age, with all its data overkill.

I stopped by to see how Owen was doing on the data from Titan. "Of the two extreme models of the atmosphere that we had," he told me, "it is beginning to look as if the low-pressure model—the all-methane model, with a surface pressure of seventeen millibars, one-point-seven percent of the Earth's at sea level, is not right. The upper-limit model—the one proposed by Donald Hunten, postulating a mostly nitrogen atmosphere—is three bars, three times the Earth's. I don't know yet what Titan's real atmospheric pressure is, but it looks like it's nearer the upper limit than the lower. We're waiting for the data from the radio-occultation experiment, in which the spacecraft as it passed behind Titan, and again as it came out from behind the satellite, transmitted to the Earth radio signals that were transmitted through the atmosphere on each limb." Refining the radio-occultation data would take considerable effort, for the scientists had to start with information from the top of Titan's atmosphere and work downward, and the man in charge of doing this, Von R. Eshleman, who is codirector of the Center for Radar Astronomy at Stanford University, did not want to be pushed into making a premature announcement that might be wrong. He was under considerable pressure to come up with a figure, not only from his fellow scientists but also from the press; indeed, at a meeting of the entire science team at eight o'clock that morning, Edward Stone, the chief scientist, had himself tried to push Eshleman into giving at least an estimate. All he would say was, "Just tell them it's a thick atmosphere."

About the only thing Owen *was* sure of at this point, he said, was something he couldn't explain: the relative darkness of the northern hemisphere of Titan—which had seemed so pronounced on pictures taken from farther out, so that there almost appeared to be a line marking the equator—had vanished as the spacecraft came farther in. Fiddle as he would with the pictures taken shortly before the encounter, he had been unable to produce the effect. He didn't know why it had disappeared any more than he could explain what had caused it in the first place. Mindful of the Schiaparelli patterns that had been such a feature of the other satellites, and which had proved in most cases to be nonexistent as the spacecraft approached them, I asked Owen if the disappearance of the contrast on Titan might mean that *that* hadn't existed, either. "Oh, no," he said. "It was clearly there farther out; it's just that closer in, it vanishes."

I decided that the best place to find out what was going on might be the morning press conference. When I arrived, Stone was giving the schedule of events for the rest of the day: The spacecraft would spend much of the time photographing the underside of the rings—at 1:58, for example, the scientists would receive a series of eight photos of the big gap in the A ring, the second biggest after the Cassini division. At 2:16 would come the closest approach to Tethys—quite far away, about a quarter of a million miles from the spacecraft, at right angles to its trajectory on the other side of Saturn. Then, at 3:09, six photographs of the Cassini division and six of the planet would begin to arrive; close approach to Saturn, when the spacecraft would be just seventy-seven thousand miles above its cloud tops, would be at 3:45—at which time pictures would be arriving of the two small co-orbital satellites, S10 and S11. Around 5:30, seven photographs of Mimas would come in; the close approach to Mimas would itself be at 5:42, when it would be fifty-five thousand two hundred miles away. At 5:50, the spacecraft would be a hundred and twenty-five thousand miles from Enceladus; from 7:08 to 8:35 the spacecraft would pass behind Saturn in relation to the Earth—in the periods before that, and after that, the radio-occultation transmitter would be sending radio waves to the Earth first through Saturn's rings and then through Saturn's atmosphere (reversing the process on the way out). While the spacecraft was behind Saturn, any photos taken, and other data accumulated, would of course be tape-recorded for later transmission to the Earth—including shots of Saturn's innermost D ring. At 7:39, still behind the planet, the spacecraft would be one hundred thousand one hundred miles from Dione (at which time it would pass back up above the ring plane, which it had passed below at the time of the close approach to Titan); almost three hours later, at 10:21, it would be forty-four thousand seven hundred miles from Rhea, the closest approach to any body after Titan; and at 8:44 the next morning, November 13, it would be at its closest to Hyperion, five hundred and forty-six thousand miles, too far for the photographs to be useful. It would, however, take useful photographs of the next-to-last moon, Iapetus. Phoebe, the small retrograde moon that was farther out, wouldn't be photographed at all.

When Smith took the stand, he showed a picture of the planet Saturn—breathtaking in its ochers and umbers—"just to remind you it is still there." People tend to forget the planet, he said, in the excitement over the rings and the satellites—a failing, if it was one, he was guilty of himself, for having done this duty by Saturn, he proceeded to show a sensational series of pictures of the rings, including Terrile's eccentric one, and also a picture of the braided F ring—it looked like the double helix of DNA—that elicited first gasps and then applause. Then Soderblom took his place; several years younger than Smith, he had a spontaneous enthusiasm that was infectious. In order to provide the press with a sort of scorecard, and to provide a sense of the speed of the improvement of resolution in the next few hours, he showed pictures of the different moons as they were then known—some still only just emerging from the Schiaparelli stage—so that they would be easier to compare with tomorrow's pictures. The spacecraft, in the next day or so, would be moving around most of the satellites, getting for the first time relatively close views of their other sides. There was Rhea, with more and more craters beginning to emerge on it, and also some strange lines, which might be crevasses. What appeared to be a dark splotch cut like a smile across the north pole. In the latest photographs, Dione was still covered with the wispy terrain; apparently the wisps were connected all over this side of the planet, suggesting to Soderblom that their origin was probably internal. (This was the first indication I had that the old arguments that had taken place about Mars and the Moon, over whether features on their surfaces were of internal or external origin—suggested that morning by Ingersoll with respect to the winds on Saturn—would be translated now to the Saturnian moons.) There would be better views, Soderblom said, of the wispy terrain. Next came a photograph of Tethys: on the same hemisphere where the "circular feature" had been—away from the black band—was the other new feature, a gigantic trench; although its resolution was hazy today, Voyager would see it four times more clearly tomorrow. (Later, the trench was estimated to be about a thousand kilometers long, three or four kilometers deep, and fifty to a hundred kilometers wide.) The most dramatic picture of all, though, was of Mimas with its huge crater filling about a quarter of the hemisphere—for the size of the satellite it was on, it was the biggest crater of any then known in the solar system. Its huge central peak, with a crater neatly centered at its top, resembled a cannon poking out from inside a porthole; accordingly, the scientists had taken to calling Mimas the "Death Star," after a gigantic, spherical, and singularly unpleasant space station in the film *Star Wars*.

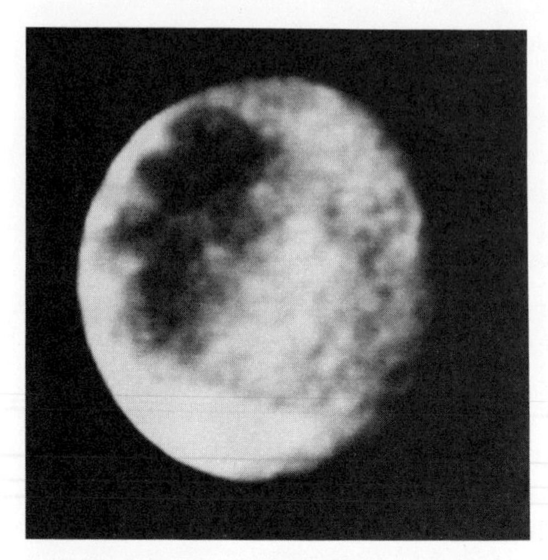

Two photos of Rhea, showing how the resolution improved on two consecutive days. ABOVE: *November 10, 1980.* BELOW: *November 11—as our own Moon might have appeared through telescopes in the seventeenth, and then perhaps the eighteenth, centuries. The rosette (above, left of center) is rapidly disappearing.*

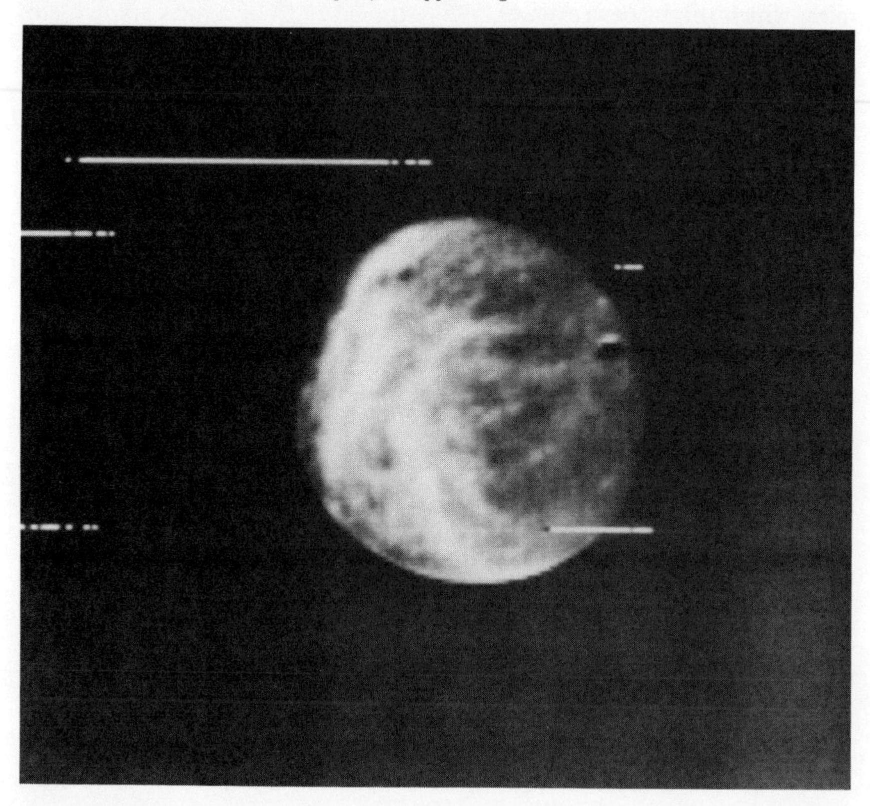

Back in one of the interactive rooms, several people—including Cuzzi, Franklin, Pollack, and Allan Cook, the head of the ring group—were looking at some new ring pictures, taken from underneath the ring plane. Having the rings held up to the light, as they were now, with the Sun on the other side, brought out much more detail. Terrile, who was wrestling with a roll of paper he was putting into the machine, said, "In this area we are looking at—one of the four big bands in the Cassini division—there are seven little bands now, where there used to be just one. That means there must be about *fifty* features in the division, which we used to think was empty."

"Maybe you could do some quick work on these rings and see if there's a satellite in there," Cuzzi said to Franklin—who, although he was one of the two men in recent times to have been most closely associated with the resonance theory, had gained a sudden interest in the alternate theory of shepherding satellites embedded in the rings; in the sudden rush of new data, everybody's ideas were in a state of flux—and, more unusual, no one seemed ashamed to acknowledge the fact.

"Do you really think there is a satellite in there?" Cook asked. Unlike the others, who were dressed informally (but neatly, in case they had to make a sudden appearance on television), Cook always dressed formally in a light gray suit with matching vest.

"I suspect there is—though I haven't found one," Franklin said. If there was, it could be so small that it would be like looking for a needle in a haystack.

"What an incredible picture this is!" Terrile said. "Shall we focus on this one area here?" He enlarged the place in the Cassini division where one ring had turned into seven—as good a place as any to look for an embedded shepherding moonlet—and pressed a button for enough hard copies for everyone in the room.

"The management is going to scream how they don't have the money to do all this processing," said Cook, who evidently was as conservative as his clothing suggested.

Nothing happened; the new roll of paper had jammed. "A fine time for the machine to break down," Terrile said, giving it a thump.

At one o'clock that afternoon, I attended a joint meeting of all the science teams. Some members of the imaging team—a handful of geologists, who knew less about physics than the astronomers—tended to avoid the joint meetings because they had little more idea than a total outsider of what the experts in magnetic fields or ultraviolet spectroscopy or radio occultation or plasma waves were saying. But even if a few members of the imaging team didn't always follow the intricacies of the other scientists' reasoning, everyone on the team admired their achievements and depended a great deal on the information provided by their instruments. Of course, all the instruments, including the cameras, worked together as part of a larger whole to understand the Saturn system, in a joint endeavor that Stone, the chief scientist of the mission, called synergy. The day before, in describing how this would apply in the case of Titan, he had said, "Imaging tells us the extent and appearance of the haze layer around Titan. Relatively high up in the atmosphere, the infrared experiment will tell us the temperatures, and the ultraviolet experiment the pressures. The radio-occultation experiment will give us the pressures farther down in the atmosphere, and may also tell us where the surface is—in other words, what Titan's radius is. The ultraviolet and infrared instruments will also tell us what some of the elements in Titan's atmosphere are, and from the temperatures and pressures we may be able to work out what some of the others are."

Stone called the meeting to order. With the encounter information from Titan in and the encounter information from the rest of the Saturn system beginning to arrive, the room was unusually full. The findings were still very preliminary: neither the ultraviolet-spectroscopy team nor the infrared-radiation team felt that their instruments had penetrated to the surface of Titan, and the radio-science team was not yet ready to talk. The head of the infrared team—Rudolf Hanel, an atmospheric physicist from Goddard Space Flight Center, in Greenbelt, Maryland—reported that the lowest temperature reading the infrared spectrometer had gotten from the center of Titan's disc, where the measurement would be from fairly deep in the atmosphere, was 65° K; clouds prevented its getting a reading from any farther down. Out at the limb—the edge of the disc—where the infrared spectrometer was looking at a cross section of the atmosphere, the highest temperature was 175° K, presumably from the aerosol layer. Hanel said he could make out the signatures of five hydrocarbons (including methane, of course) and hydrogen cyanide (HCN)—an interesting discovery, since one of the elements in that compound is nitrogen. Though it is a lethal poison, it is perhaps the simplest true organic compound. "Titan is probably the most spectacular thing we've seen," he concluded.

Hanel was followed by the leader of the ultraviolet team, Lyle Broadfoot, an atmospheric physicist with the Earth and Space Sciences Institute of the University of Southern California. Broadfoot's report confirmed the presence of nitrogen in Titan's atmosphere. Most of the scientists present seemed to feel by now that nitrogen, with all its implications for organic chemistry, was definitely its major component, in accordance with the higher-pressure model—though the precise pressure would depend on where the surface was, still unknown. (Later, it was tentatively suggested that the atmosphere was as much as 90 percent nitrogen, with less than 10 percent methane at the surface and about one percent in the colder upper atmosphere.) The nitrogen level seemed to increase as the ultraviolet instrument was moved from the center of the disc toward the limb, implying that although it might not be concentrated in the deep layer the instrument was seeing at the center, it was a common constituent of other layers.

"The ultraviolet team's finding of nitrogen fits in nicely with the infrared team's discovery of hydrogen cyanide," Owen said to me during a break. "At one time there might indeed have been a greater greenhouse effect than any we may have today, making the surface warmer than it is now—whatever that temperature may turn out to be. Now that we have confirmed the presence of nitrogen, there might be organic compounds even more complex than we thought." (During the Voyager 2 encounter nine months later, the photopolarimeter—an instrument that was not working aboard Voyager 1, and that, because it was sensitive to polarized light, could measure particle sizes—found that the aerosols in the densest orange-yellow haze layers were in the neighborhood of .5 micron, with some as small as .1 micron; Owen told me that those sizes were consistent with complex organic compounds, including polymers—long molecular chains—that might compose the aerosols.)

Owen went on: "Earlier today, when I first heard the news about the hydrogen cyanide, I called Klaus Biemann, the professor at the Massachusetts Institute of Technology who was the leader of the gas-chromatograph-mass-spectrometer team on Viking—the team that had looked for organic chemicals on Mars but found none—and he was very excited about it. He wanted to send a GCMS to Titan, one that would sample the atmosphere on the way down and that would have a drill to bore below the ground, in the event that precipitation had built up on the surface a layer of polymers." I asked Owen what else he felt should be aboard a Titan lander. "Cameras, perhaps, with lights, for it could be quite dark under those clouds," he said. "A battery of meteorology instruments and instruments for sampling the atmosphere. Perhaps it should have pontoons, in case it lands in an ammonia ocean."

Stone adjourned the meeting until five-thirty, in the hope of getting a report from the radio-science team.

The manner in which the data was trickling in, without any particular order to it—a little bit here, a little bit there, not necessarily related to other data that was coming in at the moment—was tantalizing, not to say annoying. I asked Morrison, on the way out of the meeting, if he was ever tempted just to go home and come back later—perhaps in about two or three weeks' time, when everything was assembled at JPL, all was quiet, and he and the other scientists could begin to make some sense out of it. Indeed, because so much of the action now was automatic, beyond the scientists' control, there seemed to be much to recommend this course of action—among other things, the scientists would avoid the pressure of daily press conferences. "Oh, no!" said Morrison, who seemed as caught up in the frenetic pace as anyone. "I may well come back in two weeks also, but I don't want to miss the excitement now, of *seeing* the data come in. The data will be here forever, and we'll be able to work on it forever, but there is only one moment when it's new and fresh."

The encounter was taking on the aspects of a three-ring circus. Among the official visitors to JPL who passed through the green-and-white welcome tent by the front gate were Governor Brown of California and several onetime astronauts. On my way back to the imaging-team quarters, I ran into one of the official visitors, a man in his eighties who had come from Kentucky to see the mission; he was one of the few people presently at JPL who remembered seeing Halley's Comet on its last visit in 1910. The fact that Bruce Murray, the director of JPL, had reportedly petitioned Ronald Reagan, then President-Elect (he had defeated President Carter at the polls about a week before the encounter) for funds for the proposed NASA mission to intercept the comet on its next pass in 1984, may have had to do with why the elderly gentleman had been made an official guest. When I asked him how big the comet had looked, he held his hands about three feet apart and said, "This big," as though he were describing a fish that got away. (The comet would, too: Reagan never provided funds for the mission.)

Inside the imaging-team quarters, a small crowd—guests, both invited and uninvited—stood looking up at the closed-circuit television set, on which Harold Masursky, who was being interviewed, held up several new photos of Dione, with craters now in addition to its wispy terrain. The spacecraft was already beginning to move around Dione. "Every time we change longitude on one of these moons, we see things that are totally new," he said. "It won't take much enhancement of these photos to say what the wispy stuff is."

In the first interactive room, an astronomer—Joseph Veverka of Cornell—was enhancing the photo of S11, the trailing co-orbital satellite. "It looks like a cratered baked potato," someone in the room said. It looked, in fact, like a cratered *boiled* potato, for it was whitish—clearly made of ice; it had a thin black line running across it, which puzzled everyone.

In the other interactive room, a group, going over the latest disc pack, was studying a new photo of Mimas, which, in addition to its giant crater, was now clearly covered all over with smaller ones. The scientists switched to a picture of Tethys, which also proved to be covered with

S11, the trailing co-orbital, with the black line running across it. Imaged November 12, 1980.

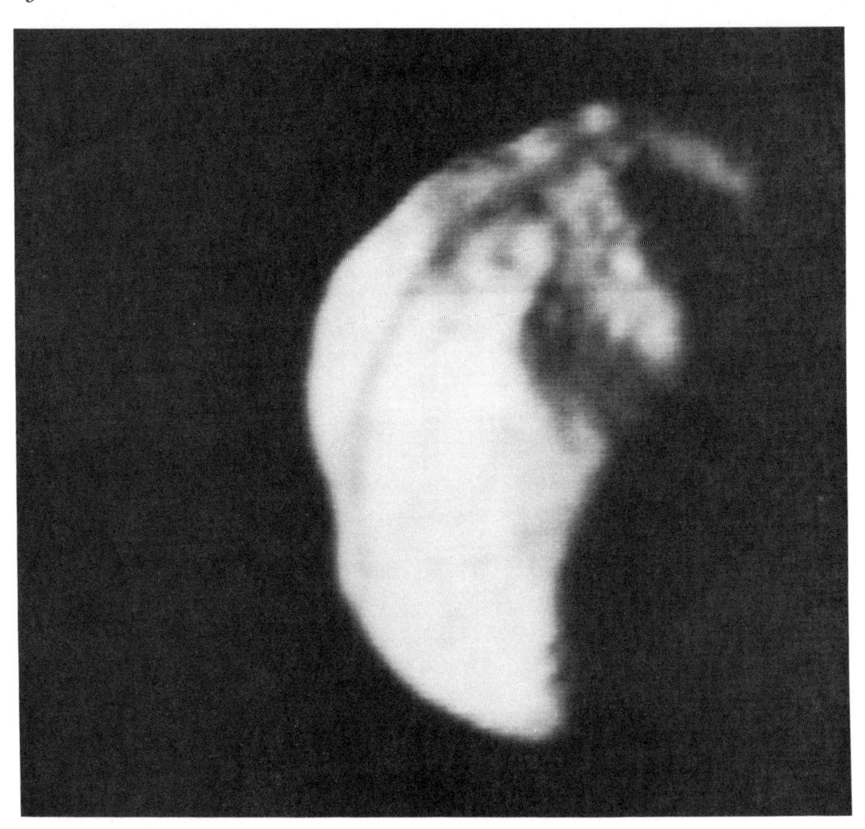

craters. The big broad north-south black strip first noticed the day before was more pronounced now—unlike the equatorial line on Titan, which vanished when Voyager came closer. Tethys had also developed an equatorial band of its own—fainter than the north-south one, but just as puzzling. Soderblom, who was standing in the back of the room, said he didn't know what the bands were, though he expected the darker, at least, was of internal origin—as, he believed, was the case with the wispy terrain on Dione. Mimas, on the other hand, appeared to be cratered more evenly all the way around, implying to Soderblom that its surface had been unaltered by any internal process since the early days of the

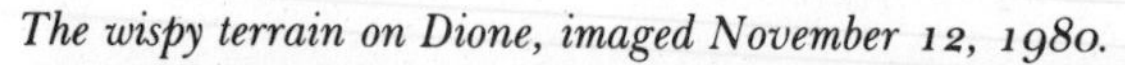

The wispy terrain on Dione, imaged November 12, 1980.

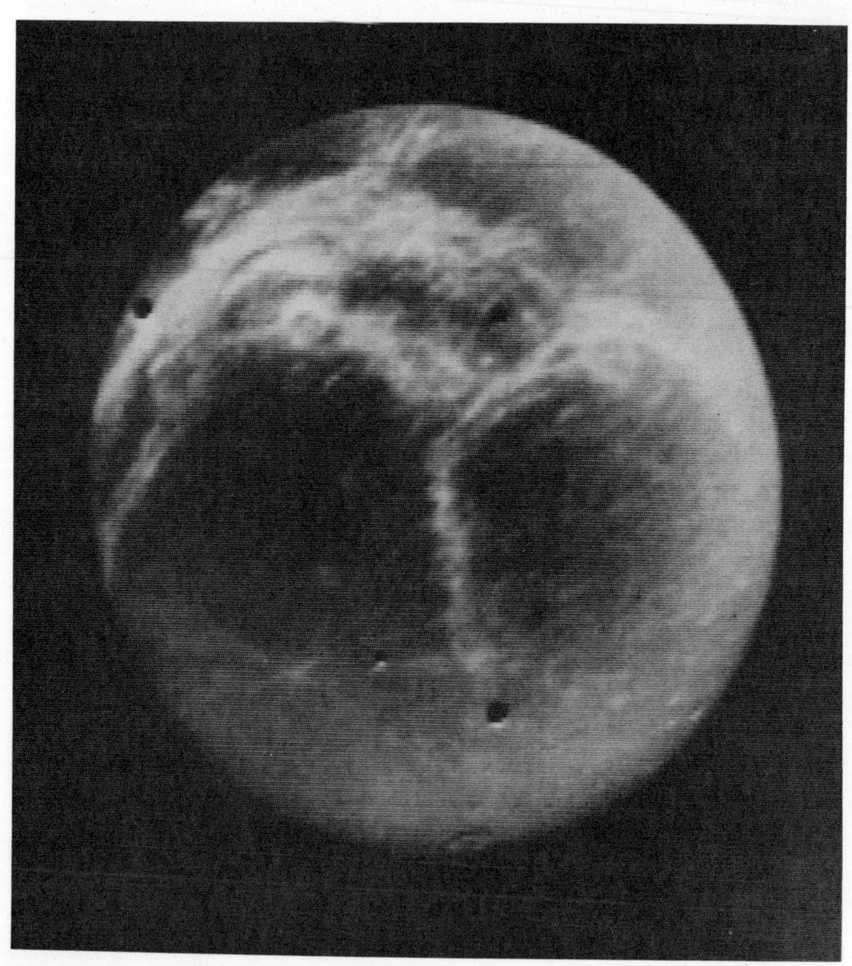

Rhea, from forty-five thousand miles, turned out to be heavily cratered. Imaged November 12, 1980.

solar system. He thought he could detect something of a pattern. "Evidently, Mimas froze entirely as soon as it was formed, for it retains the surface relief from the early days of bombardment," Soderblom said. "Dione and Tethys show some sign of change since those days. It may be that the common force that explains these differences is the rate at which these bodies accreted out of the cloud that made Saturn and its system. If a body accreted gradually, it would be a Mimas, cold and rigid, able to retain its original shape. If the body accreted rapidly, it would be a Dione or a Tethys, because there would have been more heat—and there would have been more evidence of internal activity." Soderblom's, and the other scientists', ideas were of course still changing by the minute, almost as fast as the spacecraft was changing position; a little later, Soderblom de-emphasized the accretional-heating idea; he told me that Mimas in fact very likely would have accreted *faster* than Dione or Tethys, because it was the innermost large satellite, and the accretion rate was probably greater nearer Saturn. He shifted his attention to the satellites' relative size, for a small satellite like Mimas, whose diameter was later established as 390 kilometers, would naturally lose heat more rapidly than larger satellites such as Tethys and Dione, whose diameters, respectively, were later established as 1,050 kilometers and 1,120 kilometers. The satellites' density and composition, when and if they could be determined, would naturally influence his thinking as well.

Soderblom did not know where Rhea, the next satellite out after Dione, fitted into this picture, though as it was somewhat bigger than the others (its diameter was later put at 1,530 kilometers) it could well show even more signs of internal activity. The latest pictures showed that Rhea was heavily cratered all over, though there was no clue to what caused the smudgy smile crossing the north pole. "I hope those navigation guys got all the kinks out after the Titan encounter," he said, "for as we roll over the pole, doing our seven-picture mosaic, we'll be looking right down on the smile, at one-kilometer resolution!"

In the main hallway of the imaging-team quarters, I ran into one of its members who had recently arrived, Eugene Shoemaker. In 1960, he had founded the astrogeologic studies branch of the USGS and he had been its first chief (he is still on its staff, as well as being on the faculty at Cal Tech). His field of specialty for about two decades has been the study of impact craters. He is a swarthy man with a shrewd pair of eyes and a thin mustache who looks good on television—indeed, from the earliest Apollo days, when he had trained in geology the first astronauts to go to the Moon, he had often taken to lecture stand and TV screen to do battle with other geologists, such as Masursky (whom he had hired), over whether the Moon was hot. Then as now, Shoemaker pays more attention to the possible *external* causes for features on a planet rather than the internal ones—and he has a powerful reason to do so, for during the formation of the planets and their moons, and for some time afterward, there had been a prodigious bombardment of rocky and icy bodies hurtling every which way. When I repeated to him Soderblom's remarks about internal effects on Saturn's moons, Shoemaker said he didn't disagree, but he managed to put a cold-body, impact spin on the ball: "In the earliest days, when these moons were more fluid, the craters vanished; later, as the ice hardened, they retained the craters. We can count the number of craters of different sizes on the different satellites, or on different parts of them, and then compare them with the crater distributions on our own Moon, the ages of whose different surfaces, and crater distributions, we know pretty well."

From the beginning of the solar system—when the nebula around the Sun collapsed—a little less than 4.6 billion years ago, and for the next hundred to two hundred million years, until about 4.4 billion years ago, the planets and their moons were built up relatively rapidly by the accretion of the bombardment of planetesimals and other bodies such as asteroids, meteors, and comets—some bodies were rocky, others, farther out, were predominantly icy—which had formed as the gases and other material in the nebula had cooled. Saturn and Jupiter probably formed first, their solid cores accreting within the first ten thousand years to a size large enough so that, by their gravitational attraction, they rapidly pulled in the gases from the collapsing solar nebula (before Saturn itself collapsed to its present size, it may have had a radius of about a hundred and fifty thousand miles, almost five times its present one); however, Saturn and Jupiter also continued to receive their share of the blitz of smaller bodies. Some of these bodies were in relatively circular orbits around the Sun; some, perhaps as the result of frequent encounters, had been tossed into elliptical orbits; a few were in orbits around the planets or even their moons. As the planets were built up by the impacting debris, these smaller bodies were swept up, their number reducing, at first dramatically; the blizzard, though, lasted with catastrophic intensity for perhaps the next five hundred million years or so—until perhaps 3.9 billion years ago—for, judging by our own Moon, that seems to be the oldest age of the craters we see; older craters have all been obliterated. The bombardment has been tailing off ever since, even to this day, in what planetary geologists call the "flux."

This bombardment—or rather, this population of impacting material, for there are other populations—has the largest percentage of very big bodies in relation to smaller ones (a ratio, called the size-frequency distribution, that continues to this day, though of course in vastly reduced numbers), and therefore planetary scientists can recognize its results wherever they see it by doing crater counts. Shoemaker and others feel that the blitz of large bodies that made the oldest craters visible today should have been used up well before 3.9 billion years ago—indeed, they should have been used up in the accretion of the planets prior to 4.4 billion years ago—and consequently he and other planetologists have had a theoretical problem in what they call "storing" the material in orbit for so long. Shoemaker's assumption is that the bulk of the material had formed near the orbits of Uranus and Neptune—far-out, colder regions conducive to the formation of objects composed of snow, ice, and rocks that, from the point of view of size as well as composition, would most closely resemble the comets of today (which, extending as they do in what can be regarded as a gigantic cloud stretching halfway to the nearest star, very likely are the most substantial amount of unswept-up material in the solar system now). Not only was the size and quantity of the objects, as judged by today's comets, right, but beyond that, Shoemaker told me, the celestial dynamics were such that material formed in those outer orbits at the beginning of the solar system would have been "delivered" to the inner system in the right quantities at the right time. On their swings toward the Sun, these bodies would cross the orbits of all the planets (as the trajectories of comets do still), accounting for the similarities that exist in the oldest visible craterings on our own Moon and many other moons and planets in the solar system.

At five-thirty, when the joint meeting reconvened, Stone called on Von R. Eshleman, of the radio-science team. Eshleman, a ruddy, white-haired man, gave a preliminary report on the radio-occultation experiment at Titan. As the spacecraft passed behind Titan, its radio signals were progressively distorted by the atmosphere; the radio team could thus see how the atmospheric density changed with altitude, and this measurement also gave them ratios of temperature to molecular weight. The ratios they found, happily, were consistent with a predominantly nitrogen atmosphere. The pressure for such an atmosphere, at the lowest depth for which his team had so far processed its data, was about one-third of the Earth's at sea level, over 300 millibars—and the temperature was about 65° K. Eshleman cautioned that he was probably not yet seeing Titan's surface; farther down, the pressure, and perhaps the temperature, would be higher. As for Titan's radius, that would also have to wait until the surface had been located.

Stone then asked Lyle Broadfoot if the ultraviolet team could work up an estimate of the surface pressure on Titan in time for the press conference the next morning. "See how many bars you get," Stone said.

"Right now, any bar will do," someone said. From the next rom came party noises—loud conversation, the thunks of corks popping. A number of encounter celebrations were getting under way, and the meeting adjourned. The actual moment of the Saturn encounter, some two hours earlier, had been no livelier than the Titan encounter the evening before.

On my way out, I asked Owen what he could deduce about Titan from Eshleman's remarks.

"If the temperature on the surface is as low as it now looks as if it might be, there is a possibility—indeed, a virtual certainty—of liquid-nitrogen oceans," he said. "For that, you need a temperature under seventy-seven-point-two degrees Kelvin at a pressure of about one bar, and that's what we may end up with—at least, that's how it seems at the moment. An ammonia ocean is excluded—at the temperatures we're talking about, all the ammonia would be frozen."

I asked whether liquid nitrogen was as good a solvent, with respect to organic compounds, as ammonia.

"How nitrogen behaves in liquid form, I don't know," he said; "I suspect not many people do. As a gas, nitrogen is inert, which it would be as a liquid, too—and that would be in its favor, as a solvent for organic compounds."

Back in the imaging team's own quarters—in a big room beyond the others that was used for mapping, off of which cartographers and other technicians supporting the team had their offices—the team's own encounter party was in full swing. Corks were popping. Empty champagne bottles, with daisies and irises in them, decorated a drafting table loaded with plates of potato chips and hors d'oeuvres. Adding to the gaiety was the news that Jimmy Carter, the outgoing President, had just announced that he would put money for a Venus orbital mission into the budget. Brad Smith handed me a plastic glass of champagne and asked me if I had heard about what Carter's aide, Hamilton Jordan, had said about Masursky. I hadn't. "A lady from NASA went to the White House to brief Carter on the encounter—he and Rosalynn Carter and some others were watching our closed-circuit broadcasts this afternoon," Smith said. "None of them said a word during the entire two-hour show—until Masursky came on, and then Hamilton Jordan said, 'That's the most ridiculous haircut I ever saw.' That's all any of them said, the whole time." Masursky was standing nearby, grinning broadly—and with evident pride. The party, I thought, had all the intense hilarity of a New Year's Eve party, combining, as it did, a sense of achievement with a sense of ending and concern for the future. "Some people here have been working on the program for eight years," Reta Beebe said to me, "and in one day it all roars by."

A few scientists, drinks in hand, wandered back and forth between the party and the interactives. Masursky began looking at a new picture of Dione. Its surface on the side opposite the wispy terrain was pocked with small craters. Before long, the raws stopped coming in. The spacecraft had passed behind Saturn, and the period of occultation would last for an hour and a half.

OUTBOUND

At the press conference on Thursday, November 13th, Smith began his report by saying, "I am stunned by the spectacular display of pictures we have received over the last twenty-four hours. Our amazement over one picture was distracted by the next. We've learned more about Saturn in the last week than in the entire span of recorded history." Now that so much information was in, he said, the work would soon begin a new, less speculative phase. He then showed a few slides. One was a version of the F ring that showed the weaving in and out of strands more clearly than ever; there now appeared to be three of them, though only the two outer ones seemed to intertwine, and the braiding apparently did not go all the way around the ring. He also showed a view of the spokes in the B ring—looking back toward the Sun now—in which they appeared to be lighter than the rings, and a view of the many ringlets inside the Cassini division. "Richard Terrile counted so many ringlets in the whole ring system that he got tired and stopped," Smith said. "The count will probably be somewhere between five hundred and a thousand." The existence of the D ring was finally

The hazy D ring, imaged August 21, 1981, by Voyager 2, which had a higher-resolution view of it than Voyager 1, which had discovered it nine months earlier.

confirmed; like the main rings, it, too, seemed to be divided into ringlets. Smith concluded the slide presentation with two enhanced views of S11, the tooth-shaped, trailing co-orbital moon, which looked like a fragment of something larger. "It is tempting, after seeing pictures of both co-orbitals, whose orbits are less than fifty kilometers apart, to think that they might have originally been one body," he said. Then he pointed out the thin black stripe that ran across the moon's surface; in the pictures, taken nine minutes apart, the stripe had apparently shifted position. It was thought to be the shadow of an as yet unidentified ring.

The next report was given by Laurence Soderblom. "Over the last twenty-four hours, we have seen the large icy satellites grow from the stage where they were mere names out of mythology to being real places," he said. "At last we can talk about them scientifically and draw the sorts of conclusions about them that we have drawn about places we know better, such as our own Moon." He showed some new pictures. Mimas, riddled now with craters, looked like a hunk of slag. A picture of Dione showed a number of small craters that, like the giant crater on Mimas that Masursky had shown me, had peaks in their centers—a feature of arger craters on our own rocky Moon, but here they were present on craters of much smaller size as well, and hence there were many more of them. (This had also been the case with the icy moons of Jupiter, and when I had talked to him the day before, Shoemaker had predicted it for Saturn's moons; he believed they were caused by the heat generated at impact, which melted ice around the crater walls, making a slush that sloshed back to the middle, rising up and freezing; because of the lower melting point of ice, this sort of thing would naturally happen with much smaller impacts on an icy moon than a rocky one.) Predominantly, these craters with their peaks were on the leading hemisphere of Dione; the wispy terrain, which turned out to be a network of cracks with whitened edges, was—it could be seen, now that Voyager 1 had rounded the satellite—a feature of only its trailing hemisphere. (Most of Saturn's moons—like Jupiter's and our own—are tidally locked to their parent planet, so that they rotate exactly once per orbit; this means that one face is always turned inward as they orbit, so one area is always in the lead with respect to orbital motion, and the antipodal area is always trailing.) While some scientists thought that the cracks were a result of meteoric impacts in adjacent regions—indeed, some large craters could be seen nearby—Soderblom and others still thought that Dione's interior had once been warm, and that the fractures were made by the escape of water vapor from it, a process called outgassing; the whitened edges were caused by the vapor freezing there. The same question was raised by the whitened wispy terrain on Rhea and by certain crevices on Mimas that were not whitened.

Masursky's rosette on Rhea had definitely become a large crater with rays of light-colored material strewn from it. The most heavily cratered areas of Rhea, Soderblom said, were clearly very ancient, going back to the time just after the formation of the solar system—a period of intense meteoritic bombardment. A number of moons were covered with craters packed so close together that their number approached what astrogeologists call the steady-state point: for every new crater added, an old one is destroyed, and the number per moon remains constant. In places on Rhea and Dione, the heavily cratered terrain seemed to have been flooded by material from the interior, covering the remains of the early,

Dione, following Voyager 1's encounter with it, seen from the other side. The wispy terrain, prominent earlier and now visible along the limb, turns out to be a feature of the other, trailing, hemisphere. The craters, down to relatively small sizes, have central peaks.

heavy bombardment. These resurfaced areas bore the marks of subsequent, though far fewer, impacts. Later, Soderblom suggested that dense cratering might account for the darker areas (including the smile) on Rhea and on some of the other moons that, like the color division on Titan, had become less distinct as Voyager had approached; the answer would have to wait until geologists did crater counts of the different areas—a meticulous, time-consuming task—and then matched them with the dark or light areas on the pictures taken farther out.

One moon Soderblom was reluctant to say much about was Enceladus. Voyager 1 had never come closer than 125,000 miles to it, and its closest photographs, to Davies's distress, were taken about 290,000 miles away. At that distance, the surface seemed featureless; it resembled a perfectly smooth, perfectly white snowball, opalescent against the black of space. It was in fact more reflective than any other moon in the solar system—a fact no one could yet explain. Soderblom had nothing at all to say about Hyperion, which, being less than a hundred miles in diameter and a half million miles away, was still back in the Schiaparelli phase; it was just a ball of white fuzz.

An encounter with Iapetus was still about an hour off.

After the meeting, I asked Bradford Smith how seriously he took his proposal that the two co-orbital satellites were really one that had broken apart.

"Very seriously," he said. "I'd bet on it. I'm sure that when we get better photographs in we'll see where they fit together."

I asked him if he had any ideas about retargeting Voyager 2—that spacecraft's preliminary targeting instructions had already been loaded into its computer, in case of trouble later, but in the expectation that they would be redone in light of Voyager 1's encounter with Saturn.

"Of course," he said. "But they'll have to be worked up in consultation with the rest of the team. I'd say the one area where we were caught short on Voyager One was in thinking of the rings as static features and not as dynamic ones. On Voyager Two, we'll want to retarget for the dynamic quality—we'll want more pictures of the same parts of the rings, for stereo pairs that will allow us to see in three dimensions; the more they are spaced over time, the more they will allow us to see changes."

I asked him if, barring things he hadn't known about ahead of time, there was anything he would have done differently in the last twenty-four hours.

"No, there really isn't," he said.

About a dozen people—mostly the younger ones—had stayed up all night, and most of them were still hanging around the interactives, red-eyed and disheveled. Among them were Nick Schneider and John Spencer, the two graduate students from the University of Arizona.

"We thought we'd be the elite and see it through," John said, yawning.

"There was quite a party sort of mood," said Nick, running a plaid sleeve across his eyes. "During occultation, we all went out to a pub for dinner. That put us in a good frame of mind for the rest of the evening."

"The most exciting thing for many of us was looking back at the rings from the far side, after we had crossed back above the ring plane, and finding they were bright," John said.

"When these ring pictures came in—about midnight—they were so good everyone thought they were imagining things," said Nick.

"Many people may be disappointed by the moon pictures—nothing but craters," said John.

"I disagree," said Nick. "The pictures of Iapetus that came in just a little while ago are the best so far. When they came over raw, you couldn't see anything. We all knew, though, when we put them on the interactive, they'd be interesting." He typed some numbers into one of the machines, and up on the screen came what looked like a bat-shaped object—the extreme black-and-white mottlings on Iapetus.

"At five o'clock, the pictures of Tethys came in," John said, resuming his account of the night. "They were pretty good. I waited up for those, and afterward I felt it was not worth going to bed."

Nick, however, had missed the Tethys pictures. "At four o'clock, I realized I wasn't going to make it through the night, so I went into an empty office and spent four hours crashed out on the floor under a desk. I kept waking up and worrying that someone would come in and find me. Finally, at about eight o'clock, I got up, because I could hear doors opening and closing and people moving about; I felt I wanted to get out before I had to explain to the person whose desk it was what I was doing there."

"We've sort of got our second wind," John said, yawning again. "Right now, we're looking at the last of last night's stuff on the interactive. How anxious we were to look at the moons and everything when they were small, watching them grow and grow each day on the screen! From now on, they'll be getting smaller again, receding. You won't find the same drive to look at new images on the way out as on the way in. And I don't think Voyager Two will have the same excitement as Voyager One. There's only one first time."

Nick, whose second wind seemed to have wafted a little sprightliness back into his manner, began pecking away at the keyboard to bring up the next picture on the disc pack—another one of Iapetus, blacker than the previous one because it was taken a little farther around. His fingers flew over the keys as he tried to enhance it. Looking up, he said to me, "How can you stand being here and not having your hands on the data?"

Richard Terrile, who had been up all night watching the pictures come in, was in one of the interactive rooms. "It's a real adrenaline flow to barrel into a whole flight of moons," he said. "The rings make the most delicate striations against Saturn—like brushstrokes! It makes the hair on my arms stand up."

I asked him if he had made any more discoveries.

Not on his own, he said, but he had taken part in some group discoveries. Perhaps the most startling one had been—as John Spencer had suggested—that the spokes, which had been black when the spacecraft approached the rings, were white now that the spacecraft was moving away from Saturn. "One of the models we had earlier was that somehow something dark was imprinted on the rings," he said. "That was when the Sun was behind Voyager, and the light from the rings was backscattered. Now, with Voyager above the rings again and on the other side of Saturn, the Sun is back-lighting the planet and the rings, and the light from the rings is what we call forward-scattered. Black backscattering and white forward-scattering are characteristic of particles that are very small. We are seeing the same thing in the F ring, and we think its particles are tiny, too."

(An enormous amount of data about the rings had been collected the night before by the radio-science team, which had caused a radio signal to be beamed from the spacecraft through the rings to the Earth. Though the data was still being analyzed and the preliminary results wouldn't be announced for a few days, Terrile told me later that the team was verifying, by analyzing the way in which the ring particle scattered the signal, what a number of scientists had already suspected—that the observed differences in the rings were due primarily to the number and the size of the particles they contained. In the A ring, their typical diameter was ten meters, and in the C ring two meters, while the ones in the F rings indeed proved to be far smaller: micron-sized, on the order of a few wavelengths of light. The B ring—where the spokes

ABOVE: *Before the encounter, with the Sun behind Voyager 1 so that the light from the rings is backscattered, the spokes show up as black.* BELOW: *After the encounter, with the spacecraft above the ring plane again but the Sun now on the other side of the planet so that the light from the rings is forward-scattered, the spokes are white.*

are—did not let the radio signal pass through it, very likely because the number of particles was too great, so that no estimate of its typical particle size could be made.)

"But back to the spokes," Terrile went on. "The speculation is that they're clouds of tiny particles—micron-sized, far smaller than the B ring's other particles—detached from the ring and levitated above it. Since they *are* so fine, we figure we wouldn't be able to see them unless they *were* levitated. That would fit in with the fact that the spokes are most conspicuous about a third of the way in from the Cassini division, where the B ring is optically most dense—presumably, that is where there would be the most collisions, resulting in the finest particles. All this, of course, changes the model. We needed another explanation; no one was satisfied with the idea that the spokes were somehow imprinted on the rings. But it complicates things. What is raising up the spoke particles?"

In one of the interactive rooms, Owen was sitting with some of the atmosphere people, looking at what he called the debris-photo of Titan, taken after the Titan encounter when the Sun was in back of the planet, so that Titan was a crescent, with the limb brilliant in forward-scatter; if there were any particles in orbit around Titan—a thin ring, for example, or a moonlet—it would show up best in the forward-scattered light. He searched the environs of Titan and declared it ringless, moonless, and otherwise devoid of debris. He switched to another backlit picture of Titan; this one, a little farther along the spacecraft's trajectory, so that the crescent was thinner and the shadowed part of the planet was much larger, contained a surprise: the atmosphere in the shadowed side was lit up almost all the way around the limb, making a ring that went three-quarters of the way around Titan, as though the bright crescent were reaching out to embrace the dark side. "It should be black there," Owen said, beginning to enhance. "It means that there is considerable forward-scattering caused by tiny particles in the haze layer over the aerosol layer, to reflect the light around like that. Maybe we can tell something about the size of the particles and their vertical distribution." He stretched some more, but soon reached the limits of what the machine could do. He would have to wait until the Image Processing Lab enhanced it further.

"There's a sequence of ring pictures coming up on the raws," someone announced, looking at the television monitor overhead.

Andy Ingersoll, one of the all-nighters, who had been sitting quietly in back of the room, got up, yawned, and said, "I'm going to miss them if I stay, and I'm going to miss them if I go, so I guess I'll go home." The diehards were beginning to thin out. The remaining scientists moved a little closer to the interactive.

As the scientists mostly went home quite early that afternoon, to catch up on their sleep, I went to see Richard Laeser, the mission director; he looked exhausted and ebullient, and seemed to be operating on a mixture of nervous energy and exhilaration. There had been a tense moment, he said, after the close approach to Saturn, when the star tracker had had to change from Miaplacidus to another guide star, Vega. Earlier, when the engineers had biased the trackers, to accommodate the apparent shift in the position of the guide stars as a result of the distance the spacecraft had traveled, they had found that they were unable to bias the tracker far enough to reach Vega. After some ground testing, the flight controllers concluded that after the Jupiter flyby, a small piece of plastic on a plate for a transistor lead, about the size of a pencil point, had deteriorated, causing a short circuit in a wire; and, on that assumption, they figured a way around the problem. "If we were right, we knew that with the fix we would have a margin of half a degree when we acquired Vega with that tracker for the first time after the encounter. That was all the leeway we had; and if we missed, we would lose a lot of science on the outbound leg—a lot of ring pictures, and perhaps some moon pictures, including the Rhea mosaic. But when we came out from behind Saturn, the tracker was bang on the star!"

There was another tense moment a little later, when the spacecraft recrossed Saturn's equatorial plane at what was called the Dione clear zone of the E ring—or, rather, in the thinner extension of the E ring, out to the orbit of Rhea. Because Dione has a somewhat eccentric orbit, it clears a very wide path through the E ring, 5,520 kilometers across, but because the spacecraft was coming up from beneath the ring plane at a thirty-seven-degree angle, it had to hit a point 1,700 kilometers on either side of its center. "By golly, we got through, even though, because of the previous uncertainties, we had used up fifteen hundred of those seventeen hundred kilometers," he said. "Even if we had hit the E ring out there where it is so thin, nothing might have happened; however, we have a saying around here, that we don't like to kill a spacecraft, except in the most conservative way."

The next morning, Friday, November 14th, a few early risers were working the interactives. Owen was looking at wide-angle shots of Titan from different distances away from the satellite; it was easier to get high contrasts with wide-angle than high-resolution photography, and he wanted to see if he could pin down the distance in toward Titan at which the division between its two hemispheres disappeared. When I asked him if there had been any excitement overnight, he said, "No, or at least I hope not. We're just catching our breaths." In the other interactive room, John Spencer, who said he felt better now that he had caught up on his sleep, was looking at an encounter picture of Iapetus—from this particular angle, the markings didn't look like a bat anymore; rather a large S curve divided the hemisphere into light and dark halves. "You can really see that dark patch grow as you go around the moon. We're still trying to understand why one side's black and one side's white. Is it external or internal?" Spencer said, peering into the inky black side. "There really *is* something in there, in the shadow," he went on. He pressed some buttons to increase the enhancement, but whatever it was was now gone. "Probably just a reflectivity feature," he said, resignedly. "Anyway, I like the yin-yang look of it."

The yin-yang look of Iapetus.

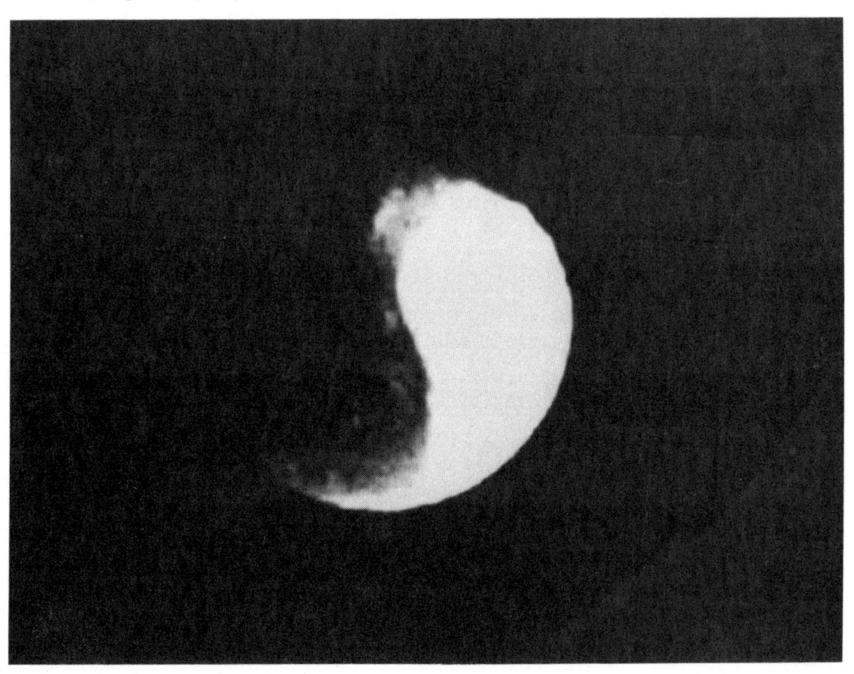

Across the hall from the interactives, Soderblom was standing in the doorway of Masursky's office. "Īo and Tēthys are British pronunciations; Ĭo and Tĕthys are European," Masursky was saying. "So both are okay on TV."

"How do you know?" Soderblom asked.

"Because I called my secretary at the Astrogeologic Studies Branch and she called a Greek scholar at the University of Arizona's classics department," Masursky said. "The Hellenic way would be Ĭo. I guess I would prefer to go with the Hellenes." Soderblom agreed that he would, too, and disappeared back to his own office.

"When we were working on names for Mars, I talked to classics professors at Harvard, Yale, and Princeton, not only for pronunciation, but for form," Masursky said as I came in. "Was it Chasma Borealis or Borealis Chasma? Some people here feel they know their Greek, and at times I've found a real resistance here to calling classics professors—but I feel if they're doing something in our area, they should call us, and vice versa."

I asked Masursky what he had been doing over the last twenty-four hours, and he told me that while he was waiting for moon pictures to come in he had been compiling lists of names for any features that might be discovered on them, for the nomenclature committee to choose from. "We didn't do as well as we might have on the Jupiter names," he said. "We wound up with too many that were difficult to pronounce and spell. English-speaking people are generally happier with Latin names than with Greek ones. Personally, I hate the name Tethys. Owen likes Greek names—he's always trying to sneak them in. I've got some ammunition, though. Look what came in this morning!" He handed me a photocopy of a glossary of Latin names from the *Aeneid*. "A classics professor at the University of Arizona sent it to me." (Later, when I remarked to Owen that Masursky had armed himself in this fashion, Owen said he wasn't worried; he had glossaries of both the *Iliad* and the *Odyssey*.)

Brad Smith walked up to Masursky, and said, "I heard you say on ABC-TV last night that Rhea had so many craters it might have been hit repeatedly with a ball-peen hammer [the kind with a rounded end that makes little dents in metal]. I thought *you* had been hit on the head with a ball-peen hammer."

Masursky said that that had not been the case.

On Friday morning, when I arrived at the press conference, Smith was just starting his talk, as he frequently did, with a spectacular photo of Saturn—it was the same one in which Terrile had noticed, in black and white, that the planet could be seen through the wispy rings, though now the photograph had been further enhanced in color by the imaging processing lab and revealed subtle differences in various parts of the rings: brushing over the creamy yellow-brown of Saturn, the A ring was ocher and the B ring almost gold. The rings' shadow ran across the planet, with the Cassini division projected brightly between two darker bands just below Saturn's equator; in turn, Saturn's own shadow ran across the rings where they soared, swirling, out of one corner of the picture; what particularly interested Smith was that the light from the rings reflected onto the dark side of Saturn like moonlight—only in this case, he said, the light was from hundreds of billions of tiny moons.

Smith was followed by Shoemaker, who had wanted to give his thoughts after preliminary observations of five of the moons of Saturn. He began with the huge crater on Mimas, a third the diameter of the satellite itself; the impact was close to the size that would have broken the moon to pieces. It would be interesting, he said, to see what the back side of Mimas, antipodal to the giant crater, was like, which it was possible to do now that the spacecraft had passed Mimas and could look back at it; on the next slide, Mimas's back side turned out to be unusually saturated with craters and in particular some big fractures. He moved on to Tethys. First he showed a picture of the huge, bifurcated trench, which covered about a third of one face of the planet. The reverse side, heavily cratered, was also the location of the evanescent round feature that had intrigued the scientists when Tethys was still in the Schiaparelli stage, and that Shoemaker now suggested might be the remnants of a huge impact that, as with Mimas, almost smashed Tethys apart, and which had cracked the far side. (If so, this seemed to explain why so many of the satellites, from far away or from close up, appeared to be characterized by some single, huge feature.) On Dione, Shoemaker pointed out a large crater in the midst of some of the wispy terrain. Shoemaker, on another slide, showed how the wispy terrain ran around in back of Dione, becoming associated with some deeper grooves. "Once again, I'm suggesting the planet was cracked," Shoemaker said. He pointed out that frequently the side opposite the big impact was relatively devoid of craters—as though they had been smeared by shaking. Certain moons, such as the theoretical parent body of the two co-orbitals, as Smith had suggested, may in fact have been cracked in two; if they were large enough, which the co-orbitals were not, the gravity of the pieces would bring them back together again. Indeed, Shoemaker suggested later that Tethys and Mimas might even have been blasted into several pieces that reassembled under their own gravity, something that could have happened several times, perhaps five times in the case of Mimas; the present trench on Tethys and the giant crater on Mimas would therefore be evidence of later impacts not quite big enough to shatter these moons yet again. Shoemaker had had this idea in the train, on his way back to Flagstaff after the encounter; there, he surprised his colleagues at the USGS by lining up some rocks of different sizes on a hillside and blasting them with bullets; some of the smaller rocks shattered (presumably in space the fragments would have remained separate), some a little larger broke into two or three pieces (in space, presumably, they would have reassembled), and some, larger still, hadn't shattered at all but were pocked with a giant crater (like Mimas—whose last big impact was not quite enough to shatter it yet again). Whether a moon was likely to shatter depended not only on its size in relation to the impact, but on its distance from Saturn, which tended to pull objects gravitationally toward itself, thereby focusing the bombardment on the nearer moons, such as Mimas and Tethys; Shoemaker calculated that the cratering rate at Mimas, 116,962 miles away, was twenty times as great as at Iapetus, over three and a half million miles away; indeed, it was by counting the number of craters on the much larger Iapetus big enough to smash Mimas and then extrapolating, that Shoemaker estimated the number of times Mimas might have been broken up. (Shattering a moon, of course, would cause it to lose any remnant of accretional heating, as well as any record of previous bombardment. It would resume its spherical shape because—in addition to a few large chunks—there would be enough small pieces to round it off; indeed, the reaccretion process would create a blanket of debris, further smoothing any rough edges.) Much larger bodies, or bodies much farther out, on the other hand, such as Rhea or Iapetus, exhibited no direct evidence, such as gigantic craters or massive trenches, of having been almost broken apart; though later Shoemaker found indirect evidence in the form of some small craters on Rhea, of an older population type, on some younger resurfaced terrain, which he felt could only be explained if that moon, too, had been smashed, sweeping up some of the smaller pieces afterward.

In the imaging team's quarters, Soderblom was taking exception to certain statements by Shoemaker, which seemed to him to skew matters too much in the direction of external impacts. "Several days ago, I thought that the wispy terrain on Dione was a sign of internal processes, and I still think so," he said. "Maybe Shoemaker is right, and the fractures on one side were made by a giant impact on the other—it wouldn't be surprising, for we see the same sort of thing on Mercury, where what we call the Weird Terrain, a rough-and-tumble hodgepodge, is antipodal to a giant impact basin—but if a meteor cracked Dione, then something leaked out, and if something leaked out, it means you had leaky stuff inside. On Dione, it also could be that the fractures got pushed open by liquid coming up from inside and freezing. On Mimas, on the other hand, the fractures on the side antipodal to the giant impact have no brightness to them. That must mean that there is not enough heat inside Mimas to have emissions or outgassing to cause the whitening; that fits in with the fact that Mimas is a much smaller body than Dione, and a smaller body loses its heat faster than a larger one; hence you hit it, and nothing leaks out. I don't agree with Shoemaker's supposition that these icy moons have rocky cores; to the extent that they have rock at all, how do you get the rock at the center and the ice on the outside? And I don't believe the greater number of craters with central peaks has anything to do with the lower melting point of ice, as Shoemaker has been saying; these moons are so cold, and the heat generated by an impact would dissipate so fast, that the ice wouldn't be melted." He felt that on both rocky and icy moons the central peaks are caused by rebounding. He put down the larger number of central peaks to a variety of factors, including the lower viscosity of ice and the lower gravity of these moons, allowing the material to rebound more.

I asked Soderblom, who seemed somewhat out of sorts, if he was having any other disagreements.

"Yes," he said. "I don't agree with Masursky's ball-peen-hammer model for the craters on Rhea."

A few hours later, I spotted Terrile in one of the interactive rooms, and I asked him, as had now become second nature, whether he had made any more discoveries.

"I've found a new ring," he said. "At least, I think I have. It's a very thin one outside the F ring. Pioneer Eleven's charged-particles experiment found evidence of a very thin ring out there somewhere, and it was tentatively labeled the G ring. I won't know for a few days whether I've found a totally new ring or just made a visual sighting of Pioneer's

G ring. For the time being, we're calling it the G ring. I don't know if it's the ring that's casting the shadow on the trailing co-orbital, and we won't know until the data is further processed, but it might be. Have you had a good look at the shadow on S-Eleven?" He pushed some buttons on the console and called up a series of pictures of which the two I had seen at the previous day's press conference were part. This series had been taken three minutes apart, and showed the narrow black stripe moving across a crater. "It's amazing that the moon's orbit and the ring are so perfectly parallel," he said. S11 might have been careering along a rail—as neat a demonstration of the perfection of celestial mechanics as anyone could ask.

I asked Terrile how he had managed to find the ring.

"I found it on a picture that was meant to bring out the faint E-ring material," he said. "When I came in early this morning, somebody told me that we had a good picture of the E ring. We weren't sure our cameras would pick it up at all. Even though the E ring is at least fifty-five thousand miles wide, it's so diffuse that it's very hard to see. I managed to detect it myself last March, when the rings were edge-on to the Earth (as they had been in nineteen sixty-six when the ring was discovered), in infrared measurements I made from Hawaii. Allan Cook and I have done a paper on it. We suggested that it might be made up of material emanating from Enceladus, whose orbit is more or less in the middle of it. Naturally, when I heard about the E-ring picture I went and called it up. Sure enough, there the E ring was, like a mist. I left it at that. Several hours later, while I was looking at some ring-plane-crossing pictures, I flipped back to that first one and made a hard copy. I left it on my desk. After a few more hours, I glanced at it and happened to notice in one corner what looked like a thin ring. I went back to the interactive and enhanced it some more. It *was* a ring."

(Much later, because of an error that had been made in the measurement of Terrile's ring, it turned out that what he had found was the G ring after all—he would have to be content with having made the first visual sighting of it. And the shadow on S11 turned out to have been cast by the F ring.)

In the other interactive room, Tobias Owen was looking at various pictures of the limb, or edge, of Titan's disc. Titan was back in the news, because at the press conference that morning the ultraviolet team had announced the discovery of two distinct layers of haze on its upper

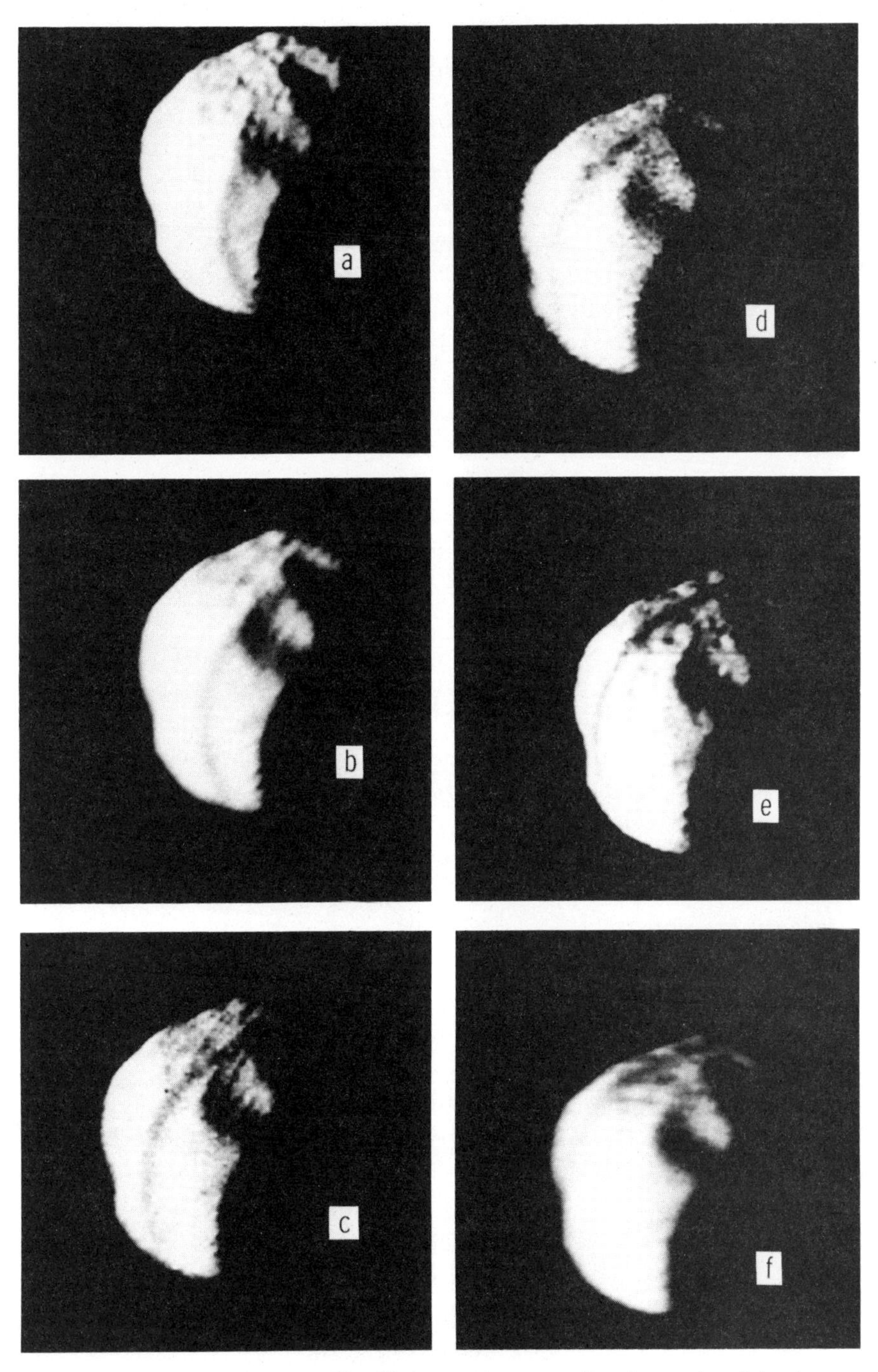

The shadow of a ring—as yet unidentified—moving across S11 in a series of six pictures imaged November 12, 1980, at three-minute intervals.

atmosphere. Owen, of course, had expected something of the sort from the backscattering on the dark side of Titan's limb, but now Smith was able to present further evidence of the layers: One of the pictures of the limb had at last been enhanced in exaggerated colors by the processing lab, which had brought out a thin blue layer of haze hovering above a second, thicker blue layer; below these two blue layers lay the far thicker aerosol layer, which gives Titan its brownish-yellow appearance. The color difference between the aerosol layer and the thin hazes above might have to do with temperature as well as density, Smith said.

"There's a problem," Owen said, pushing some buttons on the interactive. "We want to see if the two top layers we see really *are* the same as the ones the ultraviolet team has detected. The trouble is that their numbers indicate that their layers are a little higher up than ours. At some point, we will have to tie both the imaging and the ultraviolet data in with the radio-occultation data, and perhaps that will sort things out."

I asked Owen if the radio team had refined their data down to the surface of Titan yet.

"They don't know," he said. "They have got farther down now, to a level where the pressure is one and a half times the Earth's at sea level, and the temperature is ninety-two degrees Kelvin. If that *was* the surface, then the radius of Titan would be two thousand, five hundred and sixty kilometers—about one thousand, five hundred and ninety miles—and its density one point ninety-two grams per cubic centimeter. But they think they will go down a good deal farther." One thing he was sure of, he added, was that Titan had been dethroned as the largest satellite in the solar system; the level up to which the data had already been processed represented a radius somewhat smaller than the 1,640 miles of the Jovian satellite Ganymede.

In the cartography section—the room beyond the main imaging-team quarters, where the encounter party had been held two nights before—John Spencer, standing at a drafting table, was helping his professor, Robert Strom of the University of Arizona, count craters on Dione; it was to help out with such activities that Spencer and Nick Schneider had persuaded Strom they would be indispensable to the mission. Next to him, on a yellow legal pad, were several hundred notations, ~~////~~, ~~////~~, ~~////~~, and Spencer was still jotting away. Back at the University of Arizona, he said, there was a machine that helped count craters and automatically plotted their sizes—it was the only thing that would pry him away from JPL. "We're trying to see if the surface of Dione is as old as the lunar highlands, which go back before four billion years ago," Spencer said. "It's hard to count on a curvature." Across the drafting table, Jay L. Inge, a cartographer with the USGS, was drawing craters with an airbrush on a map he was making of Dione, using a Voyager photograph as a guide. "This is all very preliminary," he said. "I'm doing the transfer of the round surface onto the flat surface with the eye alone."

The door opened and in trooped about a dozen technicians from the Imaging Processing Laboratory, in the charge of a couple of the imaging-team geologists. "We thought we'd bring these people up here, so that they could see the use we make of their pictures," one of the geologists said.

Inge showed them the map he was making of Dione, and then, on a separate piece of paper, with a few quick motions of his airbrush, he drew a very serviceable crater. "Here's its rim. Here's its shadow. Here's the plateau it's situated on."

The photographic crew looked on admiringly. "How about a central peak?" one of them said.

"Here's one," said Inge, deftly making a jagged little mountain in the middle.

Impressed, one of the photographers said, "Who needs photographs anyway?"

After they had left, the geologists and some of the cartographers went to a meeting of the satellite group in the small conference room next to Smith's office. Masursky served as chairman. Strom, a medium-sized man with brownish hair, reported that the map of Dione would be ready later that afternoon; tomorrow, he and his group were going back to Arizona. They had started with Dione, he said, because it had a lower crater density than most of the other satellites, and therefore would be easier to do; the less-cratered areas had apparently been resurfaced, by material from inside the moon, late enough in its history so that the flux of meteors was greatly reduced. Strom placed transparent plastic overlays, one of the cratered surface of the ancient highlands of our own Moon and another of an ancient, heavily cratered area of the Jovian moon Callisto, on top of the still-uncompleted map of Dione; already, though, it was clear that Dione had fewer craters than either of those places—though it was *more* heavily cratered than another overlay, this one of a relatively young *mare* on our Moon. Masursky was clearly delighted. "We thought before that there were signs of internal activity on some of these satellites, but there's no question about it now that we're getting some documentation," he said.

The next morning—Saturday, November 15th—I had a hasty talk with Brad Smith, whom I caught on the run. I asked if there was anything new going on. "Yes," he said, panting up some steps on his way into the science building. "There's some interesting work being done now about the spoke problem. A number of us have thought all along that there would be no simple, brilliant statement that would make the spoke problem fall into place. By and large, the team is working well on the spokes and on other problems as well. We're moving ahead—we're beyond the stage now when we were looking at new pictures on the interactive and saying things like, 'Hey, that's a crater,' 'No, it's a reflectivity feature.' Now we're getting some real numbers. And we're beginning to find that certain nongravitational forces may be operating on the rings and spokes." Before I could find out what these were, an elevator door opened and swallowed him up.

A little later, I asked Edward Stone if he agreed with Smith that the analysis of the data had entered a new phase.

He nodded. "Normally, a scientist might get some data and then have, say, six months to think over where he is in a problem, and what he wants to look for next, before he gets his next set of data," he said. "With Voyager, the time frame has been much faster. We've known right along that we would be getting totally new data in a matter of hours, and therefore we haven't wanted to waste time thinking about a problem—unless, of course, the idea popped right out at us. The spokes are a good example of that: they were black as we approached Saturn and white as we moved away. We knew we were going to get new data from beyond the rings, and we didn't want to waste time working on the spokes until the information was all in. Right now, we're at the point where most of our information *is* in, and we can move into the phase where we are going to think more about the processes.

"Another area, besides the spokes, that has tantalized us for some time, and that we may begin to think about now, is the intertwined F ring," Stone continued. "One thing about the rings—particularly the F ring—is that they are in a radiation environment that is heavily ionizing, which undoubtedly produces a significant amount of electrical charging. We have to puzzle out whether, and to what extent, the ring particles are electrically charged, and, if they are, how they are affected by the magnetic field of Saturn as the planet rotates. Both the F ring and the spokes seem to be composed of very small particles, a few times the wavelength of light—a size easily affected by electromagnetic fields. It may be that we will have to invoke electromagnetic physics, not just gravitational physics, to explain both the spokes and the F ring."

At a meeting of the imaging team's satellite group, Soderblom produced the mosaic of Rhea, which had been assembled now and tacked to a beaverboard that he propped on top of the television set in the corner. Voyager had skimmed so quickly and so closely over Rhea's north pole that the pictures would have been smeared, except for the fact that the flight controllers came up with a maneuver they called image motion compensation, which involved turning off the Sun sensor and star tracker and shifting to an alternate guidance system, a set of three gyroscopes that provided a reference for roll, pitch, and yaw, and which, to counteract the smear, could be biased to move the entire spacecraft so delicately that even the scan platform, on which the cameras were mounted, could not duplicate the motion. Earlier, after the Titan encounter, when the flight controllers were working on the overlay to the targeting instructions already in the spacecraft's computer, they had noticed another problem: the small amount by which Titan had thrown off the trajectory was enough so that the seven photographs for the mosaic wouldn't join properly—there would be overlaps and gaps, which cartographers call "gores." They had taken an extra forty-five minutes, just sixteen hours from the Rhea encounter, to recompute the aim and the timing of the pictures. "I've seen the seven high-resolution photographs, and they have no smear," Laeser told me when I saw him a day after the encounter. "Now I can't wait to see the mosaic pasted up, to see how successful we were about eliminating gores." According to Soderblom and the others present, Laeser should be very pleased: not a white space was to be seen.

Rhea's face was a welter of ancient craters, though there were certain areas that had fewer craters, and there the craters, on the average, seemed larger. The smile was gone; the dark and light areas visible from farther out were not present on the mosaic photographs, though they were present on another photograph taken earlier that Soderblom now produced. "It looks as if the area with the fewer, larger craters is the lighter area," Soderblom said, running his hand upward over the mosaic. "And, as the spacecraft moves on over the north pole, into the darker area on that photograph, you seem to get more craters on the mosaic. I think it would be worth someone's time to figure more exactly where the dividing line on the earlier photograph falls on the mosaic. It may turn out that the color differences have to do with the density of craters, and the color differences can be synched in to the differing ages of Rhea's terrain."

Rich Terrile, who had been sitting in on the meeting of the satellite group although he wasn't a member of it, excused himself in order to go to the daily meeting of the science team and I followed him. "It ought to be an interesting meeting," he said. "With respect to the spokes on the B ring, the imaging team has taken its data about as far as it can go—it seems increasingly clear to many of us that the spokes are not the kind of structure that might be created by gravitational perturbations, which is the sort of thing we *do* understand. We need further expertise—we need the evidence of the other experiments and the knowledge of the other experimenters. Voyager is a many-faceted mission! The imaging team has been getting an awful lot of wonderful data, but then you go to these science-team meetings and you see not only are we getting this marvelous stuff, but all the other teams are, too. Also, people in different fields from our own can often see things in a different light. Any scientist, or small group of similar scientists, can get very myopic and prejudiced by their own way of thinking. All of a sudden we might get a totally new perspective by bouncing an idea off someone with a different perspective—someone, for instance, who worries about electromagnetic fields all the time, instead of just celestial mechanics, and who might see something right off that would take me weeks or months to see. There is a lot of wheel spinning at these meetings, but at least it's wheel spinning in different directions, not just the same old direction. I don't know if we've got the data that will give us the answer yet, but whatever the case, it's different kinds of thinking that's going to get us the answer. There's a variety of directions available in a place like this, and that's good."

As the science-team meetings had grown larger every day, they were now being held upstairs in a bigger conference room, whose walls were decorated with handsome posters of the Jupiter system, smaller color photographs of the Saturn system, and a couple of dozen very sprightly, very colorful paintings of Saturn by a local elementary-school class—a green egg surrounded by a blue doughnut, a purple ball surrounded by a wavy yellow halo, a brown mud pie overlapped by a larger white band.

Stone announced that today's meeting would be devoted entirely to the rings—especially the spokes. Soderblom, who had come upstairs now himself, reviewed what was known about the spokes, which was still very little. As Smith had explained to me earlier in the week, theoretically

OPPOSITE: *The mosaic of Rhea.*

the inner sections of the spokes would revolve around Saturn faster than the outer sections, and the spokes would shear apart. On that point, everyone agreed. What, then, was holding them together? Was their apparent lack of shear merely an optical illusion caused by foreshortening as a result of the angle from which Voyager had imaged them? Soderblom introduced a technician from the computer-graphics lab, who showed films of the spokes that had been reprocessed by computer; his opinion was that the lack of shear was not an illusion—though more work remained to be done on other pictures of the spokes. The real news, though, was that they seemed to be moving at close to what is called the co-rotational rate—that is, at the same rate as the rotation of Saturn itself—instead of at the orbital speeds of the rings. The implication was that the spokes were somehow affected by Saturn's magnetic field, which moves at the co-rotational rate. "We really haven't been able to solve this problem," Soderblom said when the technician had finished his presentation.

There was a silence.

Then a scientist said, "We have to assume that around the rings we really have a pretty good vacuum. Conceivably, there will be a photoelectric charging of the B ring, from sunlight."

Another scientist pointed out that the spokes seemed to be formed in Saturn's shadow, and therefore couldn't be charged photoelectrically.

There was another silence.

Someone else pointed out that, whatever the case, the particles that made up the spokes probably *were* electrostatically charged, since that was the best explanation for their hovering above the B ring.

James W. Warwick, of Radiophysics, Inc., in Boulder, Colorado, and the leader of the planetary–radio-astronomy team, said, "It could be that collisions between particles in the rings might play a part in generating some sort of electrical field that would account for the spokes." Clearly, at this stage of scientific explanations, the emphasis was on variety. The number could be narrowed down later.

At one point during the meeting, everyone's attention was suddenly riveted by a television monitor at the front of the room that had been displaying routine raws of the receding planet. The raws had been preempted by a short NASA film entitled *Coming Attractions*, which included animations of Voyager 2's upcoming encounters with Saturn, Uranus, and Neptune. Because Uranus was lying on its side in relation to the ecliptic, the spacecraft approached it pole-on; Uranus's rings were at right angles to the craft's flight path. As the spacecraft whipped past each planet in turn, gathering speed with each encounter, the scientists applauded. The animations were so tantalizing that there were boos when the film ended and a dull, colorless raw of the actual Saturn reappeared on the screen.

When the meeting broke up, Terrile, on his way out the door, said, "I'd love to see a spoke actually form! The important thing is that, whatever is going on, it's not just gravity and orbital mechanics. We still have gravity and orbital mechanics to think of, of course, but now we must also think in terms of particle interactions, magnetic interactions, and electrical interactions. That's the value of having so many types of experts under the same roof."

On Sunday, November 16th, some members of the nomenclature committee for the outer solar system met at 12:45 P.M. in a small conference room in the imaging team's quarters. The group consisted of André Brahic, of the Paris Observatory, Owen, Davies, and Masursky. Owen and Masursky did most of the talking.

Owen, as chairman, began the meeting. "All we want to do today is to establish categories of names for the features on the different satellites," he said.

"Among us, we probably have several lists," said Masursky. "Maybe we should simply associate the moons having the most features—like Rhea—with the lists containing the most names, and then we'll get home quicker."

"Maybe we should start with the moons that have the fewest features," Owen said.

"Those would be Hyperion and Enceladus," Masursky said. "We were too far away to see much of anything on them."

"Here's a myth about Enceladus himself, and it doesn't have many names, so it would do nicely for his satellite," Owen said, riffling through some glossaries. "Enceladus was a giant. Athena buried him under Sicily."

"Most of Saturn's satellites are named for his brothers and sisters, who were Titans—like Saturn himself, of course," Masursky said. "What about using the names of other giants for most of the craters on these satellites?"

"That's a thought," said Owen. "But there's one thing: after the IAU nomenclature committee's meeting in Budapest, when they approved

OVERLEAF: *Six of the photographs taken at fifteen-minute intervals on October 25, 1980, which were used in the spoke movie.*

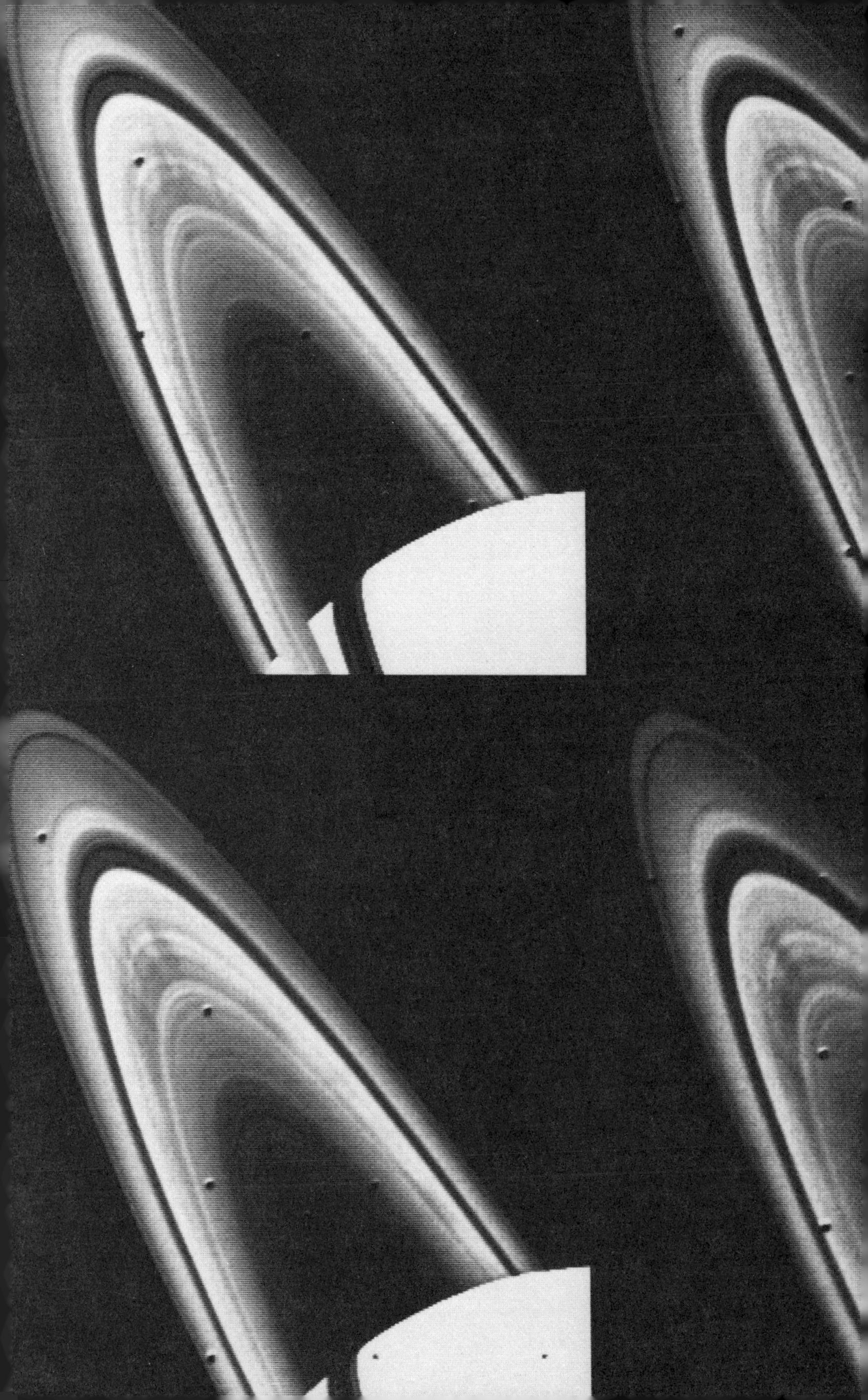

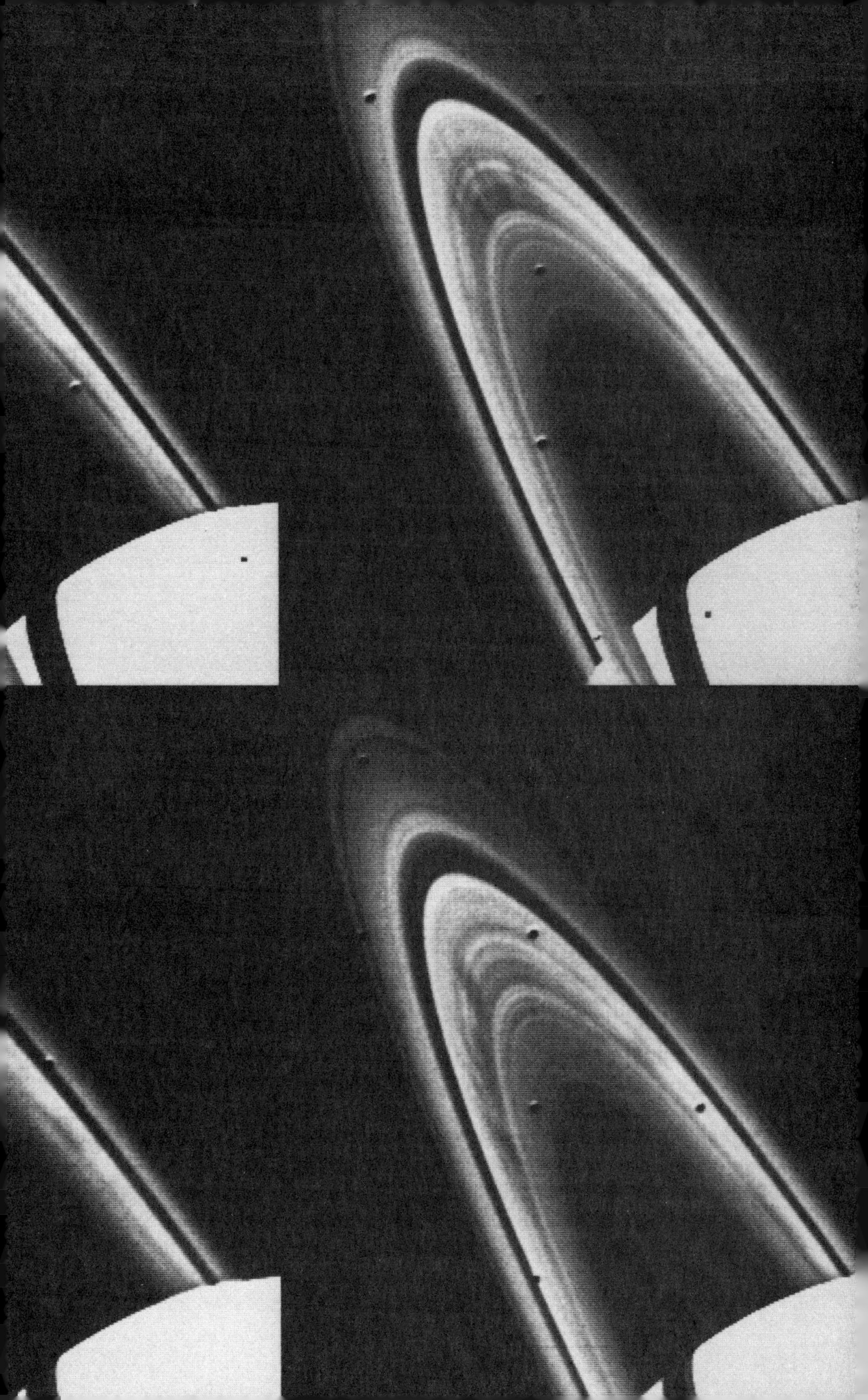

our Jupiter names, I was told that we were using too many names from Western civilization, and we should use more myths from other parts of the world—say, China or Japan."

"Maybe we should use worldwide giants," Masursky said. "What was the name of the Colossus of Rhodes?"

"I was thinking about creation myths," said Owen. "They're worldwide, and they have plenty of giants in them."

"Giant myths *and* creation myths," Masursky agreed. "I got someone at the University of Arizona to get me some names from Africa, Polynesia, Korea, and the Hawaiian Islands."

"I see a compromise that will make everyone happy," Owen said. "We'll use giant myths and creation myths, and preferentially pick Eastern names, and we'll scatter them about on all the satellites equally, so that we don't have one satellite that's, say, all Japanese."

"I wonder if there are any giants in the *Aeneid*," Masursky said, pulling out the photocopy of his glossary.

"What about the *Iliad* and the *Odyssey*?" Owen asked. "I know there are some giants in there."

"Should we combine the *Iliad* and the *Odyssey*?" Masursky said. "After all, they're one story, and if they're combined they'll take up one moon."

"I'll leave the *Iliad* out," Owen said, in a burst of magnanimity. "That way, maybe we can substitute a Japanese story—the *Tale of Genji*, or whatever. Or maybe the story of King Arthur. There are hardly any names from English legends in the solar system. I thought Iapetus would be a good place to use the Arthur legend, because Iapetus is black and white, and Arthur has so many good and evil characters. We could name a white crater for Lancelot and a black one for Mordred."

"What about the big crater on Mimas?" Davies asked.

"We could name it after Huygens, who first saw a ring around Saturn," Owen said. "Or maybe we should use his name for one of the rings."

"Are we going to name the gaps?" Brahic asked.

"I'm opposed to naming voids after people," Masursky said.

"Maybe we should put all the discoverers on one satellite—the way we did at Jupiter," said Davies.

"I like the idea of putting the discoverers of satellites on the satellites they discovered," Masursky said. "It feels nice to me."

"What about the people on the team who have discovered satellites, like Rich Terrile?" Davies asked. "He's still alive, so we aren't allowed to name anything for him."

"We'll give him a list of names and ask him to pick one," said Masursky.

"Let's get back to Enceladus, the giant Athena buried on Sicily," said Owen. "We could take Greek names for Sicilian towns and spread them all over that satellite."

"Syracuse," said Davies.

"Palermo," said Owen.

"When we get this out, some men in dark suits will pay a call on Owen," said Masursky. "Here I have five pages of names from the *Aeneid*. Maybe it would be safer to use some of them."

After forty-five minutes, the committee members, feeling the press of other business, agreed to adjourn and meet again that evening, at a local restaurant. (Six months later, at a meeting of the full committee in Bath, England, the King Arthur legend was assigned to Mimas, the *Odyssey* to Tethys, and the *Aeneid* to Dione; creation myths from all over the world were assigned to Rhea. The giants were voted down, because the majority of them came from Western, particularly Scandinavian, tales.)

At the science team's meeting later that afternoon, Stone began by appointing the scientists who would speak at the next morning's press conference, the last of the encounter. Owen would talk about Titan. Terrile would explain the latest thinking on the spokes. Smith would announce the discovery of the ring that Terrile had found two days earlier, which was now thought to be the one casting the shadow on S11. The scientists had not yet decided whether it was the G ring, detected by Pioneer 11. Rudolf Hanel would discuss a discovery that the infrared team had just made—and which Hanel, a short physicist with short-cropped white hair, took the occasion to tell the scientists about now: Saturn appeared to be emitting somewhat less heat than Earth-based measurements had indicated—though still more heat than it got from the Sun and more, relatively, than Jupiter was radiating. In other words, it still appeared that Saturn had to have an internal heat source in addition to the residual heat from its formation. In this connection, Hanel continued, the ratio of helium to hydrogen near its surface was lower than at Jupiter and in the Sun, a finding consistent with the theory that deep inside Saturn, helium had precipitated and dropped through the hydrogen, generating heat. Ingersoll told me later that he was now more interested than ever in the idea that internal energy was the main force driving the winds.

Hanel was followed by Frederick L. Scarf, a physicist with TRW's space sciences department, in Redondo Beach, California, and the leader of the plasma-waves team. He played a tape recording, relayed from the spacecraft, of the radio waves emanating from Saturn and trapped inside Saturn's magnetic field, like microwaves inside a microwave oven. There was a whoosh, followed by a crackling noise, which Scarf said might be from lightning in the planet's atmosphere, or possibly from auroras. The signals sounded as if rain were falling on the spacecraft. Lest anyone think that these were noises made by the spacecraft itself, he assured us that such sounds had been subtracted, and he played some noises from the spacecraft itself, which had been recorded by a special electric microphone when Voyager was far away from Saturn, deep in interplanetary space. Voyager sounded a little like a truck going uphill; there was even what might have been a changing of gears, followed by a higher whine, and then a thump, which might have been a rock hitting a fender but was in fact the firing of a thruster.

Next, James Warwick reported on what he called some "extraordinarily intense radio emissions" that the planetary–radio-astronomy team had detected in the vicinity of Saturn during the twenty-four hours of closest approach. After eliminating various alternative sources, Warwick declared that the radio emissions came from the rings. "The rings almost certainly have strong electric fields, possibly because of Saturn's rotating magnetic field and the fact that the ring particles move with respect to it," he said. "What's more, the particles themselves are nonconductive—in this, they resemble snowflakes and ice crystals in our own atmosphere—and there are certainly collisions between them, so charges could build up. These are the kinds of ingredients responsible for terrestrial lightning, so perhaps there is also lightning in the rings. Of course, to *see* lightning you need an atmosphere. The rings do not have a thick atmosphere, like the Earth's, but they may have a very thin one. On the Earth, you have visible flashes as well as radio emissions, but you wouldn't have visible flashes in the rings—just radio emissions. The emissions we detected usually lasted a tenth of a second, although some were as long as three-tenths; that length is comparable to that of the radio emissions from terrestrial lightning in this frequency range. Their intensity was far greater than that of terrestrial lightning, though. The voltage was a million times as strong, so we have to be careful about our interpretation."

After the meeting, I asked David Morrison, the astronomer from the University of Hawaii, about Warwick's report.

"It was important," he said. "At first, no one knew where those huge electrical discharges were coming from. We've known about them for a day or so, and most of us thought they were too big to be coming from Saturn's atmosphere. We might be able to see them if there is an atmosphere around the rings. Such an atmosphere would be incredibly thin, and no one has detected one yet." (Much later, when the data from the ultraviolet spectrometer was refined, a very skimpy atmosphere of hydrogen was found to extend thirty to forty thousand miles above and below the main rings; its source was thought to be the ice in the rings. The estimated density of this atmosphere was six hundred atoms per cubic centimeter, which is almost nothing.)

Terrile, too, felt that Warwick's report was significant. "I'm sure there is a correlation between the lightning and the spokes," he said. If, as had been suggested, the ring particles were charged on the dark side of Saturn, perhaps by Saturn's magnetic field, then at the moment they came around to the sunny side of Saturn, photoelectrons from the Sun would cause a disturbance in them. In the B ring, it might take the form of invisible lightning across the ring for thousands of miles, and this might raise up fine dust to form the spokes. In other words, perhaps the spokes and the lightning were part of a single phenomenon. That could be related, too, to Saturn's co-rotating magnetic field. "Anyway, it's an idea," Terrile said.

The press conference on Monday, November 17th, was the biggest as well as the last, despite the fact that some scientists had already left. More would be leaving that day. Terrile's talk on the spokes went well—so well, in fact, that afterward, in the question period, at least one reporter made for himself the possible connection between the spokes and the electrical discharges, which was something that Terrile, at that point, had not felt confident enough about to suggest publicly. Smith said that in the latest picture of Saturn's innermost ring, the D ring, taken on Voyager's outbound leg, it could be plainly seen that material from that ring actually was falling onto the planet. Owen, in his talk, revised some of his ideas about Titan in view of the latest refinement of the radio-occultation data. Earlier, he had reduced his estimate of a nitrogen ocean on Titan to nitrogen lakes, on the assumption that the surface was far enough down so that the atmospheric pressure reached the three-bar level—an assumption that he still felt might turn out to be true. However, the latest evidence indicated that even at the three-bar level the temperature would be too high for nitrogen lakes, except, possibly, around the poles. In the lower atmosphere, he thought, it was cold enough for nitrogen clouds. There might even be some nitrogen rain, but it would evaporate before it reached the warmer ground level.

Although the radius of Titan was not yet established definitely, Soderblom was now able to give the sizes of the other moons to within twenty kilometers' accuracy, using information provided by, among others, Merton Davies—the fruits of his holding out for repetitive photography of the moons. Their masses were provided by the radio-science team, which, by tracking the spacecraft, was able to determine how much it had been deflected from its trajectory by the moons' gravity. With these figures, the scientists were now able to figure the density of many of the satellites, too, though again within limits because of the imprecision in the knowledge of their radii; with the exception of Titan, whose density could not be figured because its radius was not known at all, the other satellites ranged between 1.1 and 1.4 grams per cubic centimeter, plus or minus a little. (Tap water is 1.0 gram per cubic centimeter.) Soderblom was even able to give a rough indication of the ratio of rock to ice in some of the satellites: Dione had two parts rock to three parts ice, Mimas had one part rock to three parts ice, and Tethys has one part rock to four parts ice. (Rhea's proportion of rock to ice was estimated a few weeks later; it came very close to that of Dione. Soderblom told me that these ratios seemed to fit nicely with the evidence of internal geological activity on some of the moons—better, certainly, than accretion rates or sizes. Dione and Rhea, with their wispy terrain—which Soderblom, for one, thought had been caused by outgassing—both happened to have a relatively large proportion of rock, and rock, of course, would be the source of whatever radioactive heat the satellites possessed. Mimas and Tethys, whose giant crater and giant trench indicated greater rigidity, and hence less internal activity, had greater proportions of ice.) The greater component of rock explained why Dione, although the same size as Tethys, looked so different—a problem that had bothered some scientists.

Later, when I asked Soderblom what he thought was the significance of all this information, he told me that it was curious that although the densities generally increased from Mimas out to Titan, they did so in an up-and-down, saw-toothed fashion, starting with Mimas, 1.2, Tethys, 1.0, Dione, 1.45, and Rhea, 1.35—in a manner opposite to Jupiter's satellites, which, moving outward, decreased smoothly in density. Soderblom had an explanation for the sawtooth characteristic: Saturn's satellites are generally about a tenth the size of Jupiter's, so that if both sets of satellites were formed by the accretion of smaller planetesimals of roughly the same size (about a mile across), which some astronomers think likely, then it would take fewer such objects to make the satellites of Saturn, and they would therefore reflect random variations more than Jupiter's; consequently, the up-and-down pattern was representative of the composition of the planetesimals themselves and, ultimately, of the solar nebula in that part of the solar system. Soderblom also had an explanation for the general, overall increase of density going outward from Mimas to Titan, the reverse of the case at Jupiter; he said the fact was consistent with a theory that, when the planets were forming and very hot, Saturn, a third the mass of Jupiter, would have had only a tenth its luminosity and heat, and therefore the low-density ice would have condensed far closer to Saturn than to Jupiter, whose inner planets would have had a greater percentage of rock. According to Cuzzi's associate, Pollack, if Saturn had during its formation once filled the volume out to the orbit of Tethys, as it may well have done, its continued collapse would have removed—taken with it—all the heavier elements, leaving water behind as the main substance in the inner system. (This situation was made somewhat less clear by Voyager 2, after which the estimated densities of Mimas and Tethys were each raised slightly—though not enough to alter the basic sawtooth pattern. The increasing density trend out to Titan was affected somewhat more. And Voyager 2 indicated that Iapetus, beyond Titan, had a lower density than the inner moons—about 1.1—but there may have been other circumstances involved there.)

From time to time over the next few months, I checked with some of the Voyager scientists by telephone. Clearly, now the research was entering a quieter, more solitary phase, and most of the scientists had returned to their regular jobs. In mid-December, when I reached Reta Beebe at her office at New Mexico State University, in Las Cruces, she told me that she was now able to provide a rough map of Saturn's east-west wind patterns. The original idea that the winds at the equator were blowing eastward, whereas the winds immediately north and south of that were blowing to the west, seemed to be correct, and, she said, she and her colleagues were beginning to be able to "put in some numbers" regarding the speeds and latitudes of the wind patterns. At the equator, the estimate of the wind speed had risen yet again to over seventeen hundred kilometers, or about a thousand miles an hour. Above and below the equatorial zone, the wind belts seemed to alternate in both hemispheres, moving toward the poles. (The data for the southern hemisphere was much sketchier than for the northern because—with the exception of the brief interval when the spacecraft swooped below the ring plane—Voyager was above the rings, which obscured the southern regions.) From the equatorial zone, which was widest, extending thirty-five degrees north and south of the equator, the winds alternated three times to the fifty-fifth parallel north of the equator, and there was some indication that the same pattern was repeated in the south. "On Jupiter, I'd have had to have stopped at about the forty-fourth parallel," Beebe said. "It's as if I stretched Jupiter's wind profile to get Saturn's." She continued, "By the way, I should tell you that when I last saw you we thought that the low contrast in Saturn's atmosphere compared to Jupiter's was due chiefly to the thickness of the haze above Saturn's cloud decks. Now we think that the clouds themselves are so stirred up that there's a lack of definition in them—and that was not the case on Jupiter. Even if the haze were thin, Saturn would not be as striking to look at."

When I reached Tobias Owen at his office at Stony Brook, he was able to give me the radio team's final figures on Titan, which were still only approximations. Titan's radius was 1,598 miles (up a little from the previous radio-occultation figure, but still less than Ganymede's), its surface pressure was 1.6 bars, and its surface temperature was 93° K, or −292° F. Titan turned out to be half ice and half rock—the rock being its core. The mantle, the material between the core and the crust, most likely water, ammonia, and methane, was, early in Titan's history, liquid because the core gave off heat—both residual heat from the accretion process and heat from radioactive elements. As Titan gradually cooled, the mantle iced up, but part of it may still be liquid. Titan's atmosphere was now thought to be as much as 90 percent nitrogen. Though methane was estimated at 10 percent or less, the amount was clearly sufficient to account for Kuiper's observations and for the complex chemistry. "Ninety-three degrees Kelvin—the surface temperature—is very close to the melting point of methane," Owen said. "It probably exists in gaseous, solid, and liquid form. Eshleman suggests that Titan's atmosphere has reached a self-regulating threshold. Seasonal variation might raise the temperature, but that would cause the frozen methane to melt and evaporate, and the evaporation would cool the surface, so the methane vapor would freeze and precipitate out again. That should happen over and over again—a kind of thermostat." The temperature was probably relatively uniform over the whole sphere of Titan, Owen said, because the atmosphere was thick enough to keep the temperature about the same at the poles as it was at the equator. However, it would drop steadily as you went upward, reaching a low of 71° K, or −332° F, at a point nineteen miles above the surface; then it would begin to rise again, reaching a high of 175° K, or −144° F, at the top of the dense aerosol layer, which was a hundred and twenty-four miles above the surface. (No one knew where the bottom was.) About sixty miles above the aerosol layer was the blue haze layer seen by the imaging team. (The team now felt that it was one homogeneous layer, rather than two distinct ones.) It was about thirty miles thick. Above it were the two detached haze layers detected by the ultraviolet spectrometer. There was no temperature data for the blue haze layer and the ultraviolet layers.

I asked Owen what he would see if he were standing on the surface of Titan.

"Not very much," he said. "In all likelihood, the dense aerosol layer would cut out most of the sunlight. Brian Toon, an atmospheric physicist at Ames, has suggested that the light on the surface of Titan would be equivalent to a moonlit night on Earth. We could be surprised, though—there turned out to be more light than we thought under the clouds of Venus. Even if there was more light than is now anticipated, I still wouldn't be able to see very far, because I'd probably be standing in a methane-and-aerosol drizzle. (Because methane on Titan exists as liquid, solid, and vapor, it would be analogous to water on Earth.) My feet would be on a methane snowbank turned a dirty yellow, because the methane would be mixed with organics—I'd be standing on a blanket of dirty slush. And there might be a lake of liquid methane at my feet."

Several months later, following a reworking of the Voyager 1 data, the presence of another major ingredient in Titan's atmosphere, argon, was deduced because the molecular weight of the atmosphere demanded a heavier gas, and argon was the only one that would fill the bill. As a result, the estimated amount of nitrogen and methane were reduced somewhat, to accommodate a 12 percent component of argon. Owen told me later that he was delighted the argon had turned up, because it got around the problem he had been worrying about of trying to derive the nitrogen from the photodissociation—breaking apart by sunlight—of ammonia, which, because of the higher temperature at which it turns to vapor, would have implied a temperature on Titan early in its history of over 150° K. While such a high temperature was not out of the question with a greater greenhouse effect, Owen nonetheless was increasingly unhappy with the idea. Another problem, of course, had been how Titan had originally managed to pull from the nebula around Saturn such a vast quantity of atmospheric gases. The detection of the argon gave him a way around both problems, because it could only have become locked into Titan's ices directly from the nebula at extremely low temperatures, suggesting that pure nitrogen, which is very similar to argon, had gotten into the ices directly, too, along with the ammonia and the methane. (The technical name for this mixture, in which gases are imprisoned within the lattices of ice crystals, is *clathrate*.) Not only does the clathrate method allow Titan to retain more gases than do other methods, such as gravity; but also the nitrogen would have been released into the atmosphere at much lower temperatures than would be required for the vaporization and photodissociation of ammonia.

After I talked with Owen, I called Soderblom to find out if the crater-counting work had borne any fruit. He said that the geologists on the team were not only counting the craters but were working up what he called size-frequency distribution—the proportions of different-sized craters in relation to each other, to tell one population from another—for the various areas of the moons. They had concluded that, contrary to what he himself had once suggested, there was no good correlation between crater sizes and the darker or lighter coloring noted from farther out, on Rhea as well as other moons. Soderblom felt that the areas with the largest size-frequency distribution—they had a preponderance of craters from twenty to a hundred kilometers in diameter—were consistent with the record noted on other bodies of the early great bombardment (the one Shoemaker had told me about) and he called this type of cratering "Population I." Because of these craters' size-frequency distribution, he felt they clearly had been made by larger bodies orbiting the Sun—certainly comets, as Shoemaker had asserted, but they might have included other Sun-orbiting bodies as well. There was some disagreement about this; some scientists felt that the match between the Population I craters around Saturn was not as close to the analogous craters on our Moon as it might be, if the orbit-crossing comets were the major factor; though Soderblom, for one, thought that there were very likely enough objects in local orbits impacting at the same time to account for the local variations. (Not all the material in the great bombardment came from the so-called trans-Uranian orbits, swinging inward across the orbits of the other planets; much of the material, of course, came from relatively circular orbits around the Sun in the vicinity of the planet they impacted on, or even around the planet itself or its moons.) The argument, of course, continues. Robert Strom, following Voyager 2's encounter with Saturn, was arguing that *none* of the Population I craters were caused by objects, of whatever nature, that came from outside Saturn's own system, because such objects would be expected to hit preferentially on the leading hemispheres of the moons; such bodies would be drawn by the giant planet's gravity toward itself, so that they would cross the paths of all the moons as they moved forward in their orbits, and hence they should be concentrated on the leading hemispheres, like raindrops hitting a moving car, which pelt the front more often than the back. The Population I craters tended to be evenly distributed all over. Soderblom and Shoemaker reply that the largest impacts—such as the one that formed the giant crater on Mimas—would have knocked a moon out of tidal lock with Saturn, or off its axis of rotation, and that this would have happened many times, so that in the days of the great bombardment, a moon's leading hemisphere very likely kept changing. Still, Soderblom's and Shoemaker's strongest argument, that the bodies that made the Population I craters were external to the Saturn system, remains the size-distribution curve.

Another, somewhat younger, population of craters, which Soderblom calls Population II, clearly did come from inside the system, because its size-frequency distribution is smaller (it is characterized by craters between ten and twenty kilometers in diameter), and because they are evenly distributed around the moons, at a time when the leading hemisphere was very likely stable; material orbiting every which way within the Saturn system would tend to paste the moons from all sides. The bulk of these smaller bodies very likely originated because the larger Population I impacts would have hurled from the moons debris, or *ejecta*, which would have gone into orbits similar to the moons' own; over a long period (in some cases, until part of a moon had been resurfaced) these fragments would have impacted back onto the moons, causing a second population of craters, smaller in size than the first but more numerous, thus obliterating some of the earlier crater record.

There was evidence also of a third, younger population of craters, Population III, much smaller in size than the first two, which presumably come from outside the Saturn system because they are indeed concentrated on the leading hemispheres of the moons, arriving long enough after the Population I bombardment so that the moons were stable on their axis. They are typically only a few meters in diameter—too small to have been seen by Voyager 1—but, Soderblom told me, their effects were clearly seen, since they have obliterated signs of early geologic activity. Apparently, much of the wispy terrain on Dione and on Rhea that should have existed over the whole surface of these moons has been abraded away on their leading hemisphere by this rain of lesser material from outside the system. Dione's leading hemisphere is particularly bland. (On Rhea, the mixing of the white wispy material into darker ices that pervade the whole planet may have resulted in the darker mottlings seen from farther out, for the dark splotches turn out to be concentrated on the trailing hemisphere. The polar smile remains a mystery.) The typically smaller size of these later craters is, of course, consistent with the flux of meteoric bodies over time, for as the populations with the biggest bodies get swept up, the populations with the smaller bodies, predominant in any population to begin with, increase proportionally.

In mid-December, when I phoned Terrile at his office at JPL, he told me that a lot of ideas about the rings that had seemed to be good bets at the time of the encounter were looking less and less good as time went on. Before I left JPL, some scientists had thought that the F ring's intertwining might be due to some sort of electromagnetic force, but now—while not ruling out the effect of Saturn's magnetic field—Peter Goldreich and his postdoctoral assistant at Cal Tech, Nicole Borderies, were beginning to wonder if it might not be a result after all of gravitational forces. "Imagine the F ring, for a moment, as a guitar string," Terrile said. "You pluck the guitar string and you take an instantaneous snapshot of it. You will see all sorts of bends and wiggles in it—something like what we are seeing in the configuration of the F ring. Well, that's not the equilibrium configuration of the guitar string; that's just the mode it happened to be in when the snapshot was taken. We are thinking that what we are seeing in the F ring might be perturbations resulting from the gravitational forces exerted on the ring particles by the two shepherding moons—they are plucking this particular string. We know now that the orbits of both these moons are eccentric, but we don't yet know exactly what their eccentricities are. It may be that these moons are periodically coming very, very close to the ring—perhaps even intersecting it—and, by their gravitational effect, are causing variable patterns in the ring particles."

And this was not the only change in thinking, for Terrile went on, "We've looked more closely at the motion of the spokes, and it turns out that the three or four spokes we've really examined—in a later set of pictures taken on the outbound leg and made into a time lapse movie—may be shearing apart after all. We're still uncertain about this." I agreed with Terrile when he said that scientific research—particularly in the early stages, when all ideas are subject to change without notice—can be frustrating.

Finally, I called Bradford Smith at his office at the University of Arizona and asked him what he was most looking forward to seeing with Voyager 2. He said, "Enceladus, of course. And we'll want to image the cloud patterns in Saturn's atmosphere at much more frequent intervals. Many of them are so ephemeral that they don't reappear from one rotation to the next, but we need to study their movement in order to understand Saturn's weather. Then, some good high-resolution photography of the entire ring system—especially the Mimas resonance regions. Voyager Two will come closer to the rings, on the bright, upper side, with the Sun at a higher angle. And some time-lapse pictures of the F ring, to see movement of the braid. And we'll want more time-lapse pictures of the spokes on the B ring. We'll want more high-resolution pictures of the C ring and the Cassini division; they're less dense than the A and B rings, and therefore the best place to see if there are moonlets embedded in the main rings. On the dark side of Saturn, we'll want to take a long exposure to see if we can detect auroras or Jim Warwick's lightning."

It sounded to me, I said, as though he would be better off if he went up there himself. He agreed.

PART II

VOYAGER 2

AUGUST 1981

INBOUND AGAIN

I went back out to Pasadena, to the Jet Propulsion Laboratory, late in August, 1981, for Voyager 2's Saturn encounter. Although Voyager 2 was launched sixteen days ahead of Voyager 1, it followed a different trajectory and took about nine months longer—or almost exactly four years—to cover the nine-hundred-million-mile distance from the Earth to Saturn. The Voyager scientists felt that the second encounter would be, if possible, even more rewarding than the first, for while they were making new discoveries they could also build on the old ones and try to answer some of the questions that Voyager 1's observations had raised. Moreover, there was a sense of urgency this time; Voyager 2 would most likely be the last spacecraft to fly by Saturn for many years to come.

Voyager 2 would get a good look at the Saturnian moons that Voyager 1 had passed only at a great distance, particularly the three outer ones. In fact, Voyager 1 had seen Saturn's outermost moon only as a point of light; this was Phoebe, a tiny dark object a hundred miles in diameter, whose elliptical orbit, eight million miles from the planet at its farthest, represents the frontier of the Saturnian system. As Voyager

2 approached Saturn, the frontier guard herself would be on the far side of her orbit, and so would not be encountered until the spacecraft was leaving the system. Hence, Iapetus, next in line and two million two hundred thousand miles from Saturn, would be the first moon that Voyager 2 met on the way in, three days before its encounter with the planet. (Increasingly, the scientists were excited about solving the mystery of Iapetus's black-and-white markings.) Then, one day and nine hundred thousand miles before the encounter with Saturn, would come the encounter with Hyperion, which Voyager 1 had seen only as a fuzzy white blur. Thereafter, everything would be crammed into a twenty-seven-hour period spanning the encounter with the planet itself: notably, the close approaches to Enceladus—some three hundred miles in diameter, possibly geologically active, and certainly the most reflective body in the solar system—and to neighboring Tethys, slightly smaller than Iapetus, whose surface is marked by a gigantic trench. Voyager 2 would not approach Saturn's other major moons—Titan, Rhea, Dione, and Mimas—as closely as Voyager 1 had done.

Voyager 2 would come much closer to the rings than Voyager 1 had and would approach their sunlit surface at a lower angle before slipping below the ring plane, just outside the G ring, on its way out of the Saturnian system. Unlike the earlier mission, Voyager 2 would *stay* below the ring plane, so its coverage of Saturn's southern hemisphere, on the outbound leg, would be better. Inbound, the Sun would be hitting the rings more directly, since Saturn was now at a different point in its orbit. Voyager 2's ring pictures would therefore be brighter and more detailed than those of Voyager 1. The scientists were particularly looking forward to them, for many of the questions raised by Voyager 1 had to do with the rings: Why is the F ring composed of braided strands? What causes the spokes—those peculiar, radial smudges moving around the broad, sixteen-thousand-mile expanse of the B ring? What causes the many narrow gaps that subdivide the rings?

Voyager 2's encounter with Saturn itself would occur at 8:24 P.M. (P.D.T.) on Tuesday, August 25th; it would come within about sixty-three thousand miles of the planet's cloud tops—more than fourteen thousand miles closer than Voyager 1. Ten days later, with Saturn and

OPPOSITE: *A group portrait of eight of the nine previously known Saturnian moons, showing their relative sizes; only Titan, far and away the biggest, which would fill the entire page, is missing. From left to right, starting at the top, they are: Mimas and Enceladus; Tethys and Dione; Rhea and Iapetus; Hyperion and Phoebe.*

its rings and other moons far behind, Voyager 2 would pass tiny Phoebe. Unlike the earlier Voyager, which, in order to pass close to the giant moon Titan, had been set on a path that would cause it to be hurled by Saturn's gravity up and out of the plane of the solar system, Voyager 2 will remain in that plane to travel to Uranus, two and a half billion miles farther out, which it will reach in January, 1986, and then one and a half billion more miles to Neptune, which it will reach in August, 1989.

By the time of my arrival at JPL on Friday, August 21st, the mission scientists were quite busy, for Voyager 2 was only four days away from its encounter with Saturn.

I started off, as I had done before, by paying a call on Bradford Smith, the leader of the imaging team. Over eighteen thousand pictures in all were to be taken during the Saturn flyby. Despite a broad face and a drooping mustache that made him look like a laconic Arizona sheriff, Smith never totally manages to conceal his brisk, New England manner. He was sitting at his desk, in front of a large picture window that provided a view of the rooftops of JPL. I asked him how he had been since last November (fine), and what he had been doing in the interim (mostly helping to plan the imaging sequences for Voyager 2). Voyager 2's observations of Saturn had been going on since June 5th. On July 4th, its narrow-angle, high-resolution camera had begun taking the zoom movie, as it had done before; this time, though, over a period of fifty Earth days—or a hundred and sixteen Saturn days (Saturn rotates once every ten hours thirty-nine minutes and twenty-six seconds)—a picture was taken at precisely the same time during each rotation to make it easier to follow the motions of various features on the planet. When the hundred and sixteen frames were run through a movie projector, the planet, with its halo of rings, loomed larger and larger, and its cloud tops sprang to life, with the winds pushing atmospheric features—in particular the white spots, though there were brown ones too—east or west in alternating latitudinal zones, at varying rates. In one zone, over the hundred and sixteen rotations, some features pulled an entire revolution ahead of others. All the while, surrounding the softly colored planet—a light caramel flecked delicately with brown-and-white wisps and an occasional speck of red—the rings spun at the familiar slightly tipsy angle, and the outlying moons flicked gaily around in their orbits, the nearer to Saturn the faster—the whole a carousel fitter for a carnival than for the outer reaches of the solar system.

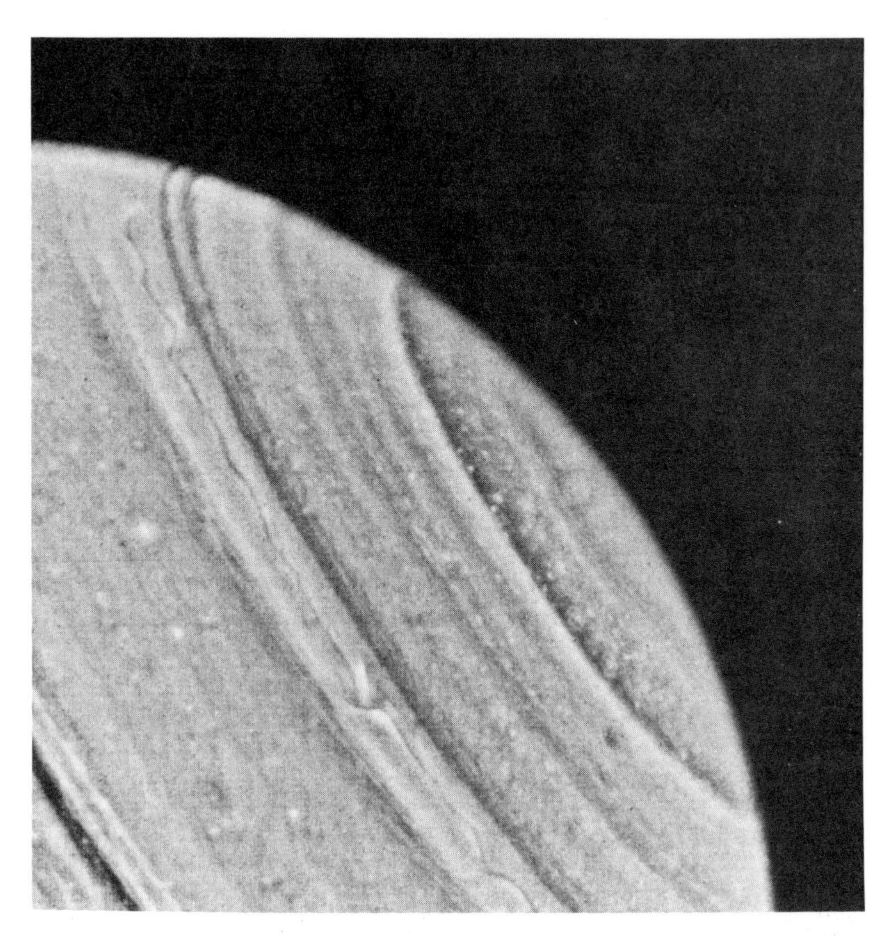

The planet appeared more active. Imaged August 15, 1981, from a distance of 6.7 million miles, through a green filter, to bring out certain details. The smallest visible feature is about sixty-two miles across.

Though it was clearly the same planet seen by Voyager 1, Saturn's appearance had changed, Smith told me. The planet now appeared to be much more active. Nine months ago, its surface features had been relatively blurred—what the scientists had called "bland"; now it more closely resembled Jupiter, with distinct spots and latitudinal bands. And the features themselves were more distinct and active than before—one, a loop that looked like the number 6, was clearly a cyclonic feature, a rotating white cloud, a wisp of which (the tail of the 6) was being pulled away by the shear of a neighboring wind current; over several days, it was pulled away altogether. (Possibly the mysterious 6, one astronomer

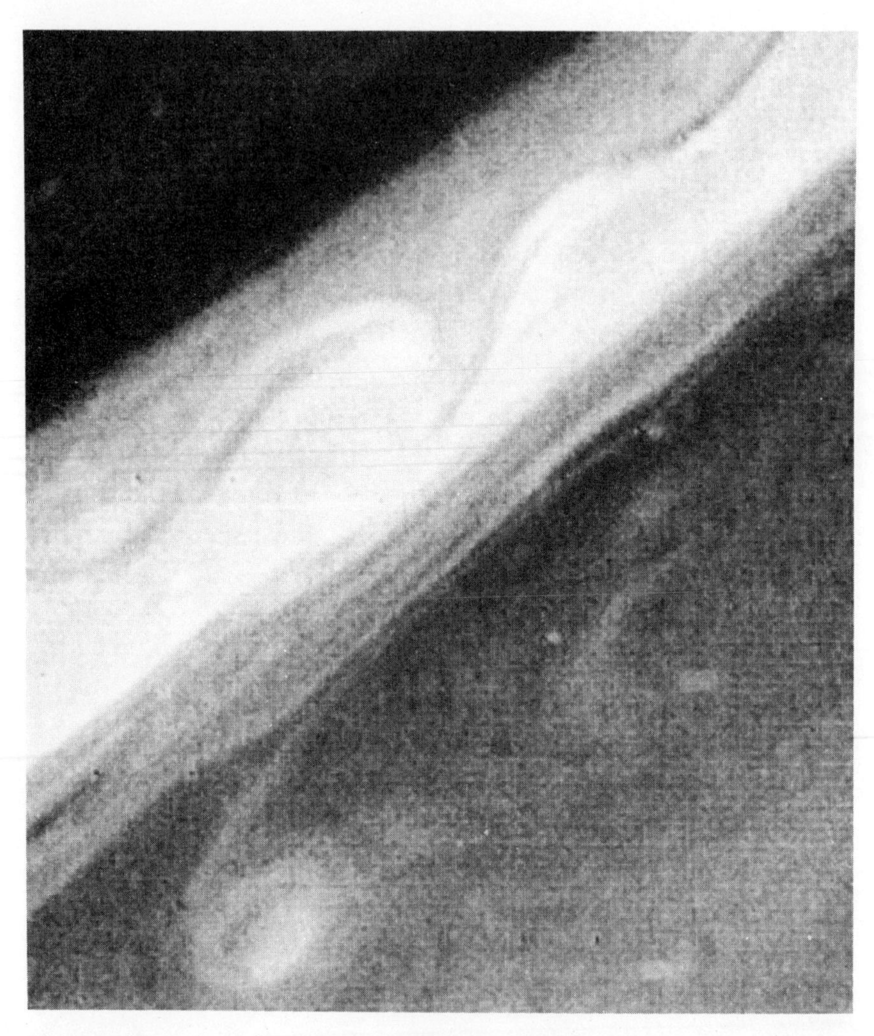

The mysterious 6, imaged August 20, 1981, when the spacecraft was four million miles away. What looks like a bracket to the north of the 6 is at the center of a westward-flowing wind zone with a velocity of two hundred and ninety miles per hour.

muttered at a team meeting, had to do with the fact that Saturn is the sixth planet from the Sun.) Nothing of the sort had been visible before. Possibly the clouds had been more stirred up then, and were less so now, making the colors and patterns more distinct. Or conceivably, the greater clarity was caused by the disappearance of a layer of haze that was thought at the time of Voyager 1's flyby to be a permanent feature of

Saturn. Smith agreed now with Reta Beebe that the haze hadn't existed after all. He pointed out a picture of Saturn, pinned to his bulletin board, that had been taken from Earth well before Voyager 1; the planet's bands were as marked on the limb, or edge, of the planet's disc as they were at its center—an indication that there wasn't any haze. If there had been, it would have been thicker and hence more obscuring toward the limb, because of the planet's curvature. "Now I'm wondering why we hadn't noticed that before," Smith said. "That picture has been around for years." He frowned, clearly annoyed with himself. "But perhaps there's more to look at because the planet really *is* more active than before, or perhaps—and this I think is the reasonable explanation—the cameras on Voyager Two are better." Even duplicates of advanced scientific instruments vary in sensitivity, he explained, and he added, "We're just very lucky if it turns out that the spacecraft with the better cameras is the one that will be going on to Uranus and Neptune."

I asked Smith if he had had any trouble with the selection of targets for the cameras. "It was tougher this time, because with one mission behind us people had more ideas about what they wanted to look at," he said. "Targeting for Voyager One was largely a matter of deciding on the most promising direction to aim the cameras." In particular, there had been a continuing debate this time between Jeffrey Cuzzi, the ring expert from NASA's Ames Research Center, who was interested in looking for small moonlets that might be embedded within the rings and might cause the narrow gaps in them; and Richard Terrile, the resident ring expert at JPL, who wanted to determine why the F ring and four of the ringlets found by Voyager 1 were slightly elliptical. (All the other rings are almost perfectly round.) Cuzzi wanted to devote a large number of pictures to a part of the B ring where he felt there was a good chance of finding a moonlet—a specific (though hypothetical) one that had come to be known as the Warwick Object, after James Warwick, who headed the planetary–radio-astronomy team, which, during Voyager 1's flyby, had detected radio discharges coming from the rings; later, Warwick and his team had determined that the signals, which reached a peak every ten hours and ten minutes, came from one particular orbital path in the B ring, where an object would orbit in that period, and thus might be caused by a moonlet there that was interacting with the ring particles. Terrile felt that that part of the B ring had already been amply covered by Voyager 1, and, furthermore, that the B ring was an unpromising place to look for elliptical ringlets, because the elliptical ringlets discovered so far were all in rather pronounced gaps, and the gaps in the B ring—the densest of the rings—were very narrow. "I'd get phone calls from those two guys from all over the country at all hours," Smith said, in a gruff manner, though it was clear that he wouldn't have wanted things otherwise. "At times, they drove me up the wall."

A little while later, when I ran into Richard Terrile outside Smith's office, I asked him about his targeting battles. He grinned and indicated things were not as bad as Smith had made out. "Cuzzi'd argue, and I'd argue, but usually we'd work things out," he said. "In the case I guess

Saturn, as seen through Voyager 2's superior imaging system on August 11, 1981, when the spacecraft was 8.6 million miles away.

Smith told you about, there was only one frame available; it had been assigned to a target that turned out to be too close to the Sun, and so it had to be retargeted. I was in Houston when the question came up. Cuzzi was anxious to put it in the Warwick area, but he knew I'd want it in the big A-ring gap, where there is an eccentric ring that from Voyager One we knew had clumps in it. Jeff called me—he had me paged in the Houston airport. I called Brad Smith, but by the time I got him it was already too late to change to the A-ring gap—so Cuzzi got it. After I got back, I managed to get another frame moved into the gap—it was one of Jeff's frames from the Warwick area—so it was a good compromise."

The last time I had seen him, Terrile had been trying to tie together the radio discharges reported by Warwick and the spokes (since both were unexplained and episodic B-ring phenomena), so I asked him now why he wasn't more interested in searching with Cuzzi for the Warwick Object, which might conceivably have something to do with the spokes. "Everyone would like to tie the spokes and the radio discharges together, but that's getting more difficult to do," Terrile said. He told me that Warwick had refined his data to the point where he could say that the orbital zone in which his suspected object lay was no more than thirty miles wide, and Terrile felt that since the spokes extended radially outward across the B ring for up to ten thousand miles they couldn't be caused by radio discharges in such a narrow area.

A few days earlier, Voyager 2's high-resolution camera had made a movie consisting of repeated four-frame shots of the rings; the object was to follow groups of spokes as they came out from behind the planet and moved around the rings. Terrile had just seen the movie at the Image Processing Laboratory, and now he took me into a small conference room adjacent to Smith's office. One wall of the room was covered with the individual frames—many hundreds of them—and because they showed the rings slanted diagonally across each frame, one had the illusion that the room was tipping. "It's hard to follow the same spokes all the way around," Terrile said. "They seem to appear, disappear, and then re-form. This goes against what we thought before. In Voyager One's pictures, the spokes had seemed to emerge strongly from Saturn's dark side, gradually disintegrating as they moved around, and all but disappearing before they went back into the shadow again. We still see them coming out of the shadow strong, and they are strongest on the leading ansa—the sweep of the rings on the left side of Saturn, in the lead as the planet moves in its orbit. They get weak, almost disappearing as they pass in front of the planet, but then they seem to form *again*, on the trailing ansa, becoming quite strong just before they go back into the shadow."

Several other members of the imaging team now entered the team quarters; the morning's joint science-team meeting, held one floor up, had let out. Torrence Johnson, a tall, trim, JPL planetary scientist with a breezy manner, came up to Terrile and said, "The radio-science team thinks we shouldn't look in the Warwick area or anywhere else in the B ring for embedded satellites." It turned out that the team, which at the time of Voyager 1's encounter with Saturn transmitted a radio beam to the Earth through Saturn's rings in an attempt to learn about their composition and structure, had reported at the meeting on the most recent refinement of those data. "They didn't find any clear gaps in the B ring, and not very many in the A ring, either," Johnson said. "And an embedded moonlet would sweep out a clear gap." The radio-science experiment was capable of finding gaps as narrow as ten to fifteen miles across, but wherever the team had looked it had found material, so the A and B rings, which the scientists regarded as a succession of many narrow ringlets and gaps, could also be viewed as continuous discs of particles in greater and lesser concentrations. What had seemed in Voyager 1's pictures to be narrow gaps might actually just be grooves, and the supposed ringlets might be ridges. "An embedded moonlet would create a gap at least ten to fifteen miles across," Johnson said. "To be certain there aren't any B-ring gaps, we'll have to wait for the results of the photopolarimeter experiment on Voyager Two, which will have a far greater resolution. But in the meantime it looks bad for the Warwick Object."

Because Iapetus—half the diameter of our own Moon—would be Voyager 2's first landfall, or moonfall, I dropped in Friday afternoon on Tobias Owen, the astronomer from the State University of New York at Stony Brook who had a special interest in that satellite. He looked very relaxed in white shorts, blue T-shirt, and a pair of white sneakers—the temperature in Pasadena was a hundred and three, and most of the scientists dressed accordingly. Owen, who had shifted his sights from Titan, which Voyager 2 would not approach as closely as Voyager 1 had done, said, "There's a lot of interest in Iapetus." "It seems to be partly covered with a very dark material, and there has been some speculation that this stuff, whatever it is, might have come from its next-door neighbor, Phoebe, which is very small and very dark." He went on to explain that any dust knocked off Phoebe by meteoric impacts would in theory form a cloud that, because of its low gravity, would not fall back to its surface but instead would gradually spiral inward toward Saturn—because Phoebe orbits Saturn in the opposite direction from all other moons—until it smacked into the leading hemisphere of Iapetus. (Iapetus, like Dione and most of the other moons, is held in rigid tidal lock by Saturn, so that one hemisphere is always in the lead as it orbits.) Iapetus's leading face is indeed black, but, puzzingly, so is its trailing face—though to a much lesser extent—and there is a sort of smear between them. The rest of Iapetus is a snowy white; the moon exhibits greater contrast in albedo—light reflectivity—than any other body in the solar system. Areas with an albedo of 50 percent (extremely bright) are next to areas of 5 percent—snow against asphalt. The only other celestial objects as dark as Phoebe and as the dark parts of Iapetus, Owen told me, are carbonaceous chondrites—meteors rich in carbon that were formed in the outer reaches of the solar system during its earliest stage, four and a half billion years ago—and others related to them.

A little later, I talked with David Morrison, the astronomer from the University of Hawaii, who is another of the imaging team's Iapetus experts. He said that Phoebe might well have begun life as a carbonaceous chondrite. Its retrograde motion around Saturn indicates that it was not formed in orbit—as the other moons presumably were, since they all move in the same direction as Saturn's rotation—but that it was probably formed elsewhere and then captured, a theory supported by the fact that its orbit is more elliptical than those of the other moons and is steeply inclined, being at an angle of about thirty degrees to Saturn's equator. The other moons (with the exception of Iapetus, whose orbit is slightly inclined—a little less than fifteen degrees) orbit along the equatorial plane. Except for Phoebe, Saturn's moons are thought to have formed where they are—to have accreted from the primordial planetary nebula, a spinning disc far broader than the present rings, which may be surviving remnants of it. If Phoebe was a captured chondrite, Morrison told me, it would be extremely ancient, predating the formation not only of the other Saturnian moons but of Saturn itself.

At the imaging team meeting later that day, Candice Hansen, a blonde, tanned young planetary scientist who had studied astronomy at the University of Arizona and who was now at JPL and who knew of Morrison's fascination with Iapetus, pitched across the room to him what she said was her latest model of the satellite—a brand-new, white softball, one of whose two interlocking figure-eight-shaped leather panels she had blackened in with India ink; Morrison himself produced it later at a meeting of the entire science team, saying it was as good a model of Iapetus as any and would have to do until something better came along.

Not everyone was convinced that the black material on Iapetus came from Phoebe. Among the doubters was Laurence Soderblom, the young, sandy-haired geologist who was the deputy leader of the imaging team and the chief of the United States Geological Survey's Branch of Astrogeologic Studies, in Flagstaff, Arizona. (Throughout the planetary program, the geologists have always seemed to be more firmly rooted to the ground than the astronomers, whose theories have often seemed more

A "hard copy" of an interactive-machine image of Iapetus, received at JPL on August 22, 1981, at 3:12 P.M. The complex boundaries between the black and the white material convinced some scientists that the black was probably of internal origin. The numbers in the margin—framing all interactive images—provide such information as catalogue number, type of exposure, and date and time (day of year, Greenwich Mean Time) when the picture was taken.

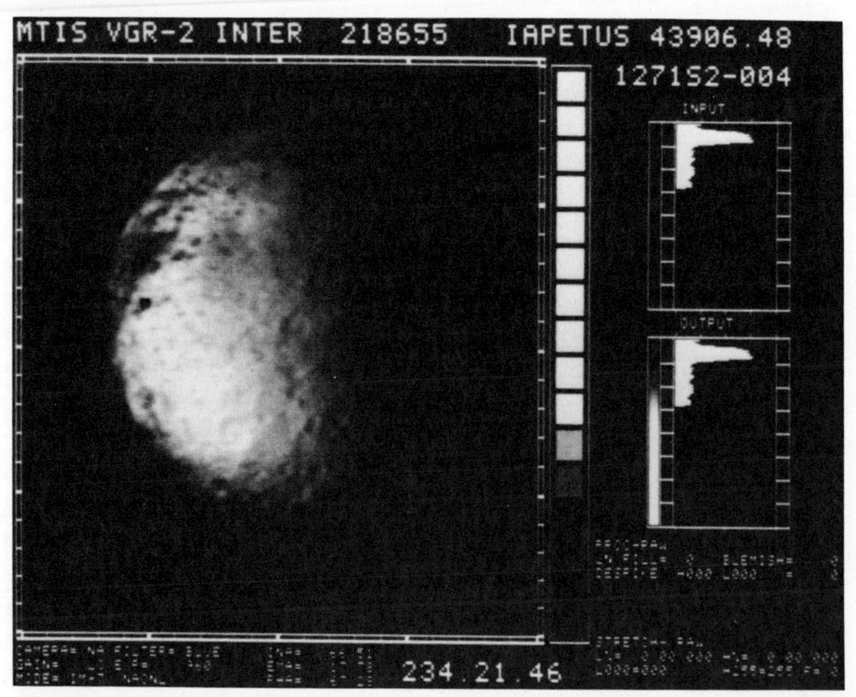

winged and ethereal.) I saw him in his office, next to Bradford Smith's, the following afternoon—Saturday, August 22nd—a few hours before the close encounter with Iapetus, which was to take place at 6:30 that night. Torrence Johnson had just come in with a couple of the most recent pictures of Iapetus, part of a series of ten taken from a distance of 738,000 miles. (The encounter that evening would be 178,000 miles closer than that.) "Now I'm convinced the black stuff didn't come from Phoebe," Soderblom said, studying the pictures. "The boundaries between the black and the white are too complex. On Dione, which we got a good look at last time and which has apparently been pelted by a rain of smallish debris from space, we see a rather uniform, grayish surface on the leading hemisphere, but the transition as you move back to the lighter material on the trailing hemisphere is a gradual, even one. On Iapetus, the boundary between the black and white is abrupt and it zigzags. We really ought to be considering models for an internal origin." He ticked off a couple of possibilities: that Iapetus was a black, perhaps carbonaceous, body covered in places with white ice or that it was a white body with black material welling up from inside and spreading out in the manner of lava flows.

Soderblom and Johnson went into the main hallway of the imaging team's quarters, where they were joined by Eugene Shoemaker. A television set suspended from the ceiling was showing the pictures of Iapetus as they came in from the spacecraft—the unprocessed raws, normally low on detail and contrast. Iapetus, of course, displayed such strong contrasts that these raws were more striking than most. The three men hurried down a small corridor and into one of two adjacent, dark, windowless rooms, which held the two interactive computer terminals. Unlike the procedure during Voyager 1's flyby, when the imaging team had had to wait until the incoming raws were collected in batches of sixty and put on a disc pack for use on the interactives—something that could take several hours—the pictures from Voyager 2 were put on a temporary disc pack as they came in and could be called up on the interactives almost immediately (an improvement that had been made, some of the scientists hinted darkly, less for their benefit than for that of JPL's public-affairs office, which planned to televise some of the scientists, live, on a third interactive set up in an adjacent building, discussing the pictures as they came in).

Soderblom called up the new series of Iapetus pictures. More of the white than the black was in view, making Iapetus look very much like our Moon, for the white, which resembled the lunar highlands, was pitted with a jumble of craters. The black areas, however, unlike the dark *maria* on our Moon (which tend to be big, round bull's-eyes), looked like ink stains that had been brushed across the white. Thus—to an untrained eye, at least—the white appeared to be showing through from underneath. The scientists, of course, could not make any such assumption.

"We'll have to do a crater count, to see if the crater density is lower on the white or on the black," Soderblom said. Counting craters—because of the flux over time in their frequency distribution—is a way of telling the relative ages of terrains: the area with fewer craters—and particularly the lower proportion of big craters—would be younger and so presumably on top. It was extremely hard, though, to make out craters in the black area. "You can see a crater rim there, poking up through the black," said Shoemaker, pointing to a tiny white circle. The scientists checked to see if any of the larger, deeper craters in the black area had white floors, which would indicate that the black material overlay a predominantly white moon; black-floored craters in the white area would, of course, indicate the opposite. Over the next few minutes, the group found both sorts of craters. All three geologists stared at the screen and at the hard copies as though they thought they could extract the answer if they only concentrated long enough. When I mentioned this impression a while later to Harold Masursky, who was himself doing a good deal of staring at images, he said, "I find that the harder I look at a picture, the more my preconceptions are reinforced."

I was in the imaging team's quarters a little after eight that night, when the first of eleven pictures taken during the close encounter with Iapetus at 6:30 was just about to arrive. (The signals from the spacecraft—from which the pictures and other data were assembled—now took an hour and twenty-six minutes, two minutes more than before, as Saturn was a little farther from the Earth, to cover at the speed of light the nine-hundred-million-mile distance to the Earth.) A few members of the team, including Tobias Owen and Carl Sagan, an astronomer from Cornell well known for his books and television appearances, were already gathered around a television monitor in one of the interactive rooms. At encounter, the spacecraft was over a different part of the satellite than it had been for the earlier pictures—farther north and almost directly over the trailing hemisphere, where there was more white. In due course, pictures would come in showing the terminator—the line where night meets day and the Sun's light rakes across the surface, showing the features in the highest relief. "Here comes Iapetus!" said Owen. The raw that flicked on the interactive screen showed, in the white area, a

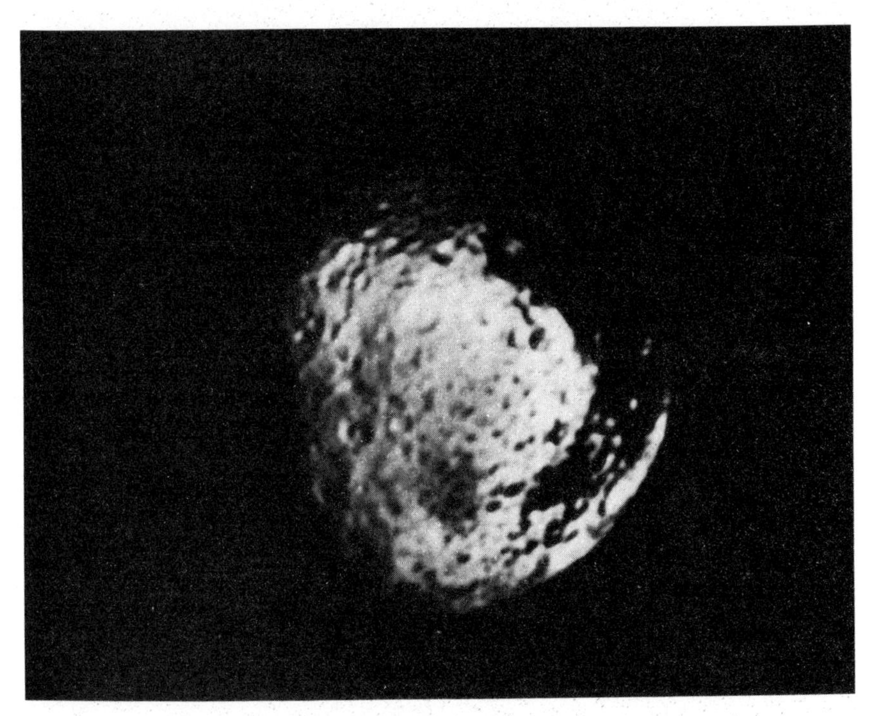

Two pictures of Iapetus, taken during the encounter period and enhanced for detail and contrast, with the tarry black in stark juxtaposition to the frosty white. Note the white crater bottoms or white central peaks of craters in the black areas, and the black bottoms in the white areas. Imaged August 22, 1981, from six hundred and eighty thousand miles away; the resolution is thirteen miles.

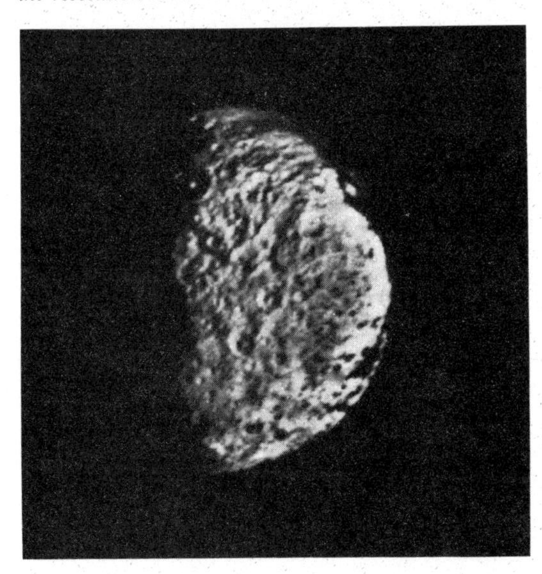

big white basin with a black rim and, in the black area, a big black crater with a white central peak rising from its floor. I asked Owen if the black material still looked to him as if it had come from Phoebe. "Yes," he said. "But I've just heard some contrary evidence. Dale Cruikshank, a colleague of David Morrison's at the University of Hawaii, has just announced the results of spectrometer studies he made that show that the spectrum of Phoebe is quite different from the spectrum of the black parts of Iapetus, and that indicates that the black material on the two moons is not the same."

Sagan, who was also fond of the Phoebe-dust theory, was not persuaded by this evidence to abandon it. He said he planned to look at the pictures of Iapetus with an eye to finding black material piled up against the crater walls that faced in the direction of the leading hemisphere—the direction of Iapetus's orbital motion—where dust might be expected to accumulate if it had come spiraling down from Phoebe. In the next few minutes, he managed to find at least one crater with what might have been dust piled up against the wall facing the leading hemisphere—though he had to admit that, against the black of the crater wall, the little black mound was difficult to make out.

Dale Cruikshank's results, which by now were already known to most members of the imaging team, were reported in detail at the joint science-team meeting on Sunday afternoon. Though both Phoebe and the dark parts of Iapetus had the same spectra as carbonaceous material, the similarity ended there; on spectrographs of the two moons, the line representing the black material on Iapetus took off steeply—in the manner, Cruikshank said, of the spectrograph of a carbonaceous chondrite that had landed in Australia, called the Murchison meteorite—while Phoebe's spectrograph gave a more or less level line, which, he said, was more like that of the Class C carbonaceous chondrites, an entirely different variety. There was a complete silence. "They're not even close!" someone said, finally. Sagan looked pensive.

Cruikshank's evidence seemed decisive to most of the scientists, though of course there were people who discounted it. After the meeting I talked to James Pollack, an astronomer from the Ames Research Center, who had been a student of Sagan's at Cornell. "Cruikshank's work is good, but there's nothing that you can't get around if you're imaginative enough," he said. "The Phoebe material would be moving on a collision course with Iapetus and so would hit it extremely fast—at two hundred and twenty miles a minute. At impact, a lot of the material would be vaporized, so you would expect to end up with stuff of a different chemical composition. Phoebe, we know from Cruikshank, is chondritic material, and if you vaporize that you can make another type of carbonaceous material, called kerogen, that would fit the spectrum of the Iapetus black." And at that day's imaging-team meeting, which was held just before the afternoon joint science-team meeting, Sagan himself, who is well known among his colleagues for sticking to his guns, had said, "If Cruikshank is right about Phoebe, there are still plenty of other external sources of chondritic material—meteors, comets—for the black on Iapetus to have come from. No, I don't regard Cruikshank's data as an obstacle at all." There was a pause, and then someone said, "Except that the source is now hypothetical."

Cruikshank's findings did convince many of the waverers. Eugene Shoemaker, who even though he is a geologist is sometimes apt to think like an astronomer, had been wavering toward the Phoebe-dust theory, but now he appeared be settling into agreement with his more down-to-earth colleagues. "I've been trying to support Sagan's argument, but it gets harder all the time," he said. "Two days ago, I'd have said there was bombardment. Now I'd have to say I'm ninety-five percent sure the stuff was erupted." And Owen, too, was changing his mind. Although later he would not rule out the idea that the dust might have come from a body other than Phoebe, as Sagan had suggested, now Owen was seriously considering the idea that the black material on Iapetus was indigenous to that moon and finding the notion just as fascinating as that of the more sensational migration from Phoebe. "It could be a kind of carbonaceous goo," he said, referring to the state in which organic compounds are thought to have been originally formed on planets. "We may be looking at a stage where carbonaceous material has not yet become a mineral; it could be mixed in with the ice."

Whatever the case, Voyager 2 had now penetrated to those colder reaches of the solar system where compounds of carbon had condensed as a solid directly from the primordial solar nebula—the vast, cold cloud of particles from which the solar system formed. The volatile materials—which include carbon compounds such as carbon dioxide or methane—had, of course, already been driven outward by the intense heat generated when the huge cloud collapsed, with the Sun coalescing hotly at its center. Carbon compounds are present on the Earth and the other inner planets either because they were bound chemically into the rocks that formed them, later to be released by volcanic activity, or because they were transported there in meteoric or cometary bodies formed farther out. The same is true of water. (Shoemaker, who with others has computed the number of cometlike bodies that hit the Earth in the great bombardment, and also the amount of water they contained, says that the total is equivalent to all the water in our oceans.) Carbon compounds and water could condense directly from the solar nebula only in the colder regions beginning with the asteroid belt, which lies between Mars and Jupiter—water condensing first, and then carbon compounds, whose freezing point is lower. Some asteroids appear to contain carbon, as also seems to be the case with Phobos and Diemos, the two tiny moons of nearby Mars—but these may well be captured asteroids. In the neighborhood of Saturn, it would have been cold enough for carbonaceous material to have condensed only in the orbits of the outer moons, since the planet's origin—from the collapse of a vast disc of particles—was itself accompanied by enormous heat. Because of the lower temperatures in the outer region of the Saturnian system—according to Owen, increasingly partial to indigenous carbon—it is quite understandable why it is present on Saturn's outer moons; one need not resort to Phoebe dust to explain it. The rings of Uranus, two and a half times as far from the Sun as is Saturn, are dark, suggesting they contain carbonaceous material. Carbonaceous matter is also found in the icy bodies of comets, which are thought to have formed beyond the orbit of Uranus. Phoebe itself, Bradford Smith later told me, may actually be a degenerate comet nucleus, captured by Saturn. If Phoebe was formed in a different part of the solar system, that would explain why its spectrum was different from that of the black material on Iapetus. "Phoebe, Iapetus, and the carbonaceous chondrites are more like comet nuclei than they are like anything else," Owen said. "They're richer in volatile materials than anything known on Earth. The universe at large may be very different from what we're used to near the Earth—the outer planets and moons may be much more typical. Iapetus is compelling even though we're seeing it at poor resolution. Like all the planetary missions, this one makes you yearn for the next."

Jeffrey Cuzzi had arrived at JPL the day before and had spent Saturday night looking over a new series of several score photographs of the A and C rings and the Cassini division, in search of the moonlets he was sure were embedded there, causing the gaps. "I'm in a jitter mode," he said, when I ran into him after Sunday's joint science-team meeting in an office he shared with Terrile and Pollack. "I had only two hours' sleep last night, and I can't figure out what to do next—whether to go back to my hotel and sleep so I can come back and look through the pictures some more tonight, or work through the day now."

There was much at stake in Cuzzi's moonlet search, of course, because its failure or success might make or break the alternative explanation for the formation of the gaps—the resonance theory, which had it that they were caused by gravitational interactions with one or another of the major moons beyond the rings.

The resonance theory, though, was still in trouble on at least two counts: first, there appeared to be more gaps than there were resonance orbits; and, second, although some gaps appeared to coincide with resonances, there was not yet any direct evidence that resonances in fact caused gaps. At the time of Voyager 1, there had been a notion that any particles left in a gap swept clear by a resonance might form a thin, elliptical ring, and that the presence of such a ring in such a gap might give credence to the theory. Now, however, some of the scientists weren't so sure the leftover particles would behave this way. One elliptical ringlet had in fact been found in a gap definitely associated with an outer moon—a gap in the inner part of the C ring that was in resonance with Titan. Nonetheless, the current explanation for elliptical rings is that they are caused not by resonances but by the gravitational effects of a couple of very small moonlets inside a gap, one on either side of the particulate material, which would both distort its circularity and keep the particles in line. The F ring—which is elliptical—has a pair of such shepherd moonlets. Peter Goldreich, professor of planetary science at the California Institute of Technology and a man who is perhaps the most productive theorist about rings and gaps, wanted to get more pictures and better measurements of the elliptical rings and their gaps, to see if they could be related to the outlying moons. Goldreich was not one of the mission scientists, but Terrile, himself the discoverer of most of these rings, was keeping in close touch with him.

I caught up with Terrile in one of the interactive rooms, where he was conducting his own search for embedded moonlets—although he was not convinced of their existence he felt he should still look for them. During Voyager 1's flyby, he had discovered S15, the moonlet that orbited four hundred miles outside the A ring, shepherding the outer edge, holding it in place by keeping its particles from dissipating outward. I asked him if he had named the moonlet yet—Owen, as head of the nomenclature committee for the outer solar system, had given him a list to choose from, including the names of several Titans and giants.

The rings as observed August 22, 1981, when Voyager 2 was 2.5 million miles away. The spokes on the B ring are particularly evident. The enormous number of features within the rings made scientists think twice about the resonance theory, which could account for only a limited number of them.

Terrile said he hadn't got around to picking one yet but was considering the name Atlas. "Atlas held up the world, sort of the way S15 seems to be holding in the A ring," he said. (Atlas was indeed Terrile's eventual choice, and the name was duly endorsed by the nomenclature committee. It was later determined that while Atlas does help keep the A ring in place, the two co-orbital satellites, though farther out, are more massive and play a larger role; nonetheless, the name remains.)

Terrile's search for embedded moonlets was a good deal more methodical than his discovery of Atlas, which had been quite serendipitous.

"So far, I've gotten through a third of the frames taken for the ring movie," he said. "Even though I'm looking at only every second frame, it's tedious work." He was concentrating on two of the largest gaps, he told me, on the assumption that if gaps were indeed caused by moonlets, then the biggest gaps should contain the biggest moonlets. One of the two gaps was at the inner edge of the Cassini division, immediately outside the B ring; the gap included an orbital path that was thought to be in a two-to-one resonance with Mimas—but as the gap itself was some three hundred miles wide, substantially wider than the Mimas

The rings, observed August 23, 1981, when the spacecraft was two million miles away, exhibited an even greater amount of structure. The limb of Saturn, ghostlike, is visible through the rings; the bright stripe around the planet is sunlight passing through the Cassini division, the darkest (and hence most transparent) part of the rings.

resonance alone could account for, it seemed a good place to look for a moonlet; should one turn up, it would presumably be credited with holding the outer edge of the B ring in place. The other gap that Terrile was studying—a wide one in the C ring—contained not only the thin elliptical ring but what might possibly be a couple of small, shepherding moonlets.

"Ha! This might be a moonlet!" he said, pointing to a picture on the interactive's screen that seemed to show a speck in the C-ring gap. He made a hard copy of that picture and of the next one in the series; if the speck was a moonlet, its position in the next picture ought to have changed, but if it hadn't moved—or, worse, if it disappeared—the speck would be simply a blemish (or a blem, as the imaging team called such things). He put the first picture on top of the second, being careful to align them exactly, and then rapidly flipped one side of the top picture up and down, so that if there was any movement it would stand out, as in a film. "It's a blem," he said.

To rest his eyes, Terrile every once in a while would take a break from moonlet search to see what was new elsewhere. Now he called up on the interactive a picture of the outer edge of the A ring with, twenty-seven hundred miles beyond it, the thin circlet of the F ring, a white streak against the black of space. Voyager 2's pictures were full of surprises, for many of the pictures seemed to run counter to observations or theories popular at the time of Voyager 1; one such surprise was the F ring, which at the time of the earlier encounter had appeared to be kinky and composed of several separate ringlets wrapped around each other in a helical manner that some astronomers called braided, but which now appeared to be composed of only one strand, with clumps, but no kinks or braids. Although Voyager 2 was still two and a half million miles away from the F ring, Voyager 1 had been at that distance when its high-resolution camera picked up the kinks and braids. One explanation for the kinks had been offered at the time by Peter Goldreich and Nicole Borderies, who felt the kinks were caused by the two shepherd moons, one on each side of the ring. Though the moons were small—each was estimated to be only about a hundred and twenty-five miles in diameter—their gravity should set up perturbations among the micron-sized particles that made up the F ring, and the perturbations should be particularly great whenever the inner moon overtook the outer moon and passed it. After Voyager 1's encounter, Terrile had likened the F ring to a guitar string just after it had been plucked. Voyager 2's narrow-angle camera had very nearly caught the passing of the two moons—

Voyager 2's narrow-angle camera shot the F ring—straight and unkinked, unlike Voyager 1's view of it—almost as the two shepherding satellites were passing each other. Imagery made August 15, 1981, when the spacecraft was 6.6 million miles from Saturn.

the plucking; yet when it was aimed at the point on the ring where the shepherd moons had come abreast, there was no sign of any recent disturbance—kinks, braids, clumps, or whatever. Terrile, though, was expecting better data on the F ring to come in during the next couple of days, in the period just before and just after the close encounter with Saturn, when five separate sections of the ring would be photographed two or three times each, from different angles, so that the sections could be viewed stereoscopically; this would give the imaging team an idea of exactly what the F-ring structure was.

Even now, though, Terrile was not without a discovery for long. In the picture on the interactive screen, the inner shepherd moon could just be made out; he enhanced and enlarged it so that the tiny satellite changed from a small round dot to a slightly larger oval. "Here is the first indication that the inner shepherd is not round but oblong," he said. "We're at the stage now where every picture will have something new in it. Some people thought Voyager Two would be an anticlimax after Voyager One, but the whole mission—it truly is a single enterprise—just goes on getting better and better." Terrile's attention was suddenly riveted by a number of white specks that, for an instant, he thought might be a whole flock of orbiting moonlets. As they didn't show up on the next picture, he declared them a gaggle of blems and returned to his labors among the gaps.

When Bradford Smith called Sunday's imaging-team meeting to order, almost everyone was present, crowded around the long table in the meeting room. No one's attention span seemed very long. Terrile, sitting to Smith's left, kept tossing a ball in the air, which he said was the latest model of Iapetus—instead of a softball with one figure-eight section blackened, it was an ordinary, sectionless rubber ball (yellow, as a white one had been unavailable) with a long black smudge going halfway around it, each end of which was rounded. Another team member, G. Edward Danielson of the California Institute of Technology, rushed in and showed Smith the latest image of Hyperion, the first to show that satellite as other than a round blob. Smith, who was talking with Soderblom on his left, looked surprised and passed the picture on to Morrison, on his right, who said, "Wow!" Then Smith resumed his conversation.

Andrew Ingersoll, the leader of the imaging team's planetary-atmosphere group, kept watching a television screen at the rear of the room, where a series of raws of high-resolution pictures of different features on Saturn itself were coming in, frequently rushing from the room and returning with a hard copy, which he would pass to an associate, Anne Bunker, a planetary scientist at JPL. Once he returned with a red apple—a model, apparently, only of itself, for he took a large bite out of it. As it was lunchtime, many of the scientists were picking nervously at takeout orders from the JPL cafeteria, mostly soggy lettuce with inedible-looking flaps of processed cheese; only Sagan looked as though he were eating something good: he was nibbling at an elegant little salad that had apparently come, in a tiny black shopping bag, from a French delicatessen in downtown Pasadena.

While the scientists ate, Danielson showed a Voyager 1 picture of Tethys—moving outward from Saturn, the third major satellite after Mimas and Enceladus; it showed the prominent black band horizontally across it. Danielson said that a colleague had suggested the black band might be some sort of shadow, possibly (as the satellite was near the ring plane of Saturn) from one of Saturn's rings—in which case one would expect the shadow to move as Tethys orbited the planet, which it didn't.

"But if you had material around Tethys, then you'd expect the shadow in the same place," Danielson concluded.

Masursky, never one to be content with shadows in lieu of substance, said, "Are you suggesting that there is a ring around Tethys?"

Danielson, who didn't really want to commit himself to the suggestion, said it was a question that ought to be considered.

There was a babble of disapproval as the scientists packed away the remains of their lunches. (The band turned out not to be a shadow but was very likely the true color of the planet—whiter frost may have been distributed in the polar regions.) Clearly the scientists were in the situation one of them had described, at the time of Voyager 1, of just letting the new material wash over them and thinking up questions that ought to be asked. Science—at the beginning of an investigation, at least, before lines harden—is, simply, inquiry, and a planetary flyby mission, transient as a shooting star, may be one of the purest demonstrations of the fact.

A particularly sensational raw of Saturn appeared on the screen, and Ingersoll darted out of the room, not to return.

I tracked Ingersoll to his office, which he shared with other members of the planetary-atmosphere group—it was beyond the two interactive rooms, at the end of the L-shaped corridor—and asked him what he was up to. He looked up from his desk, where he had been sorting through a welter of prints from the interactives, and said, "Today, we'll be getting pictures of five or six specific atmospheric features—a few brown spots and a white one. The trouble is, we had to do our targeting six weeks ago, in mid-July, and the spots all have different rates of motion. Voyager One's pictures of atmospheric features on Jupiter were so clear from so much farther out that, before we did our Jupiter targeting, we had months and months to learn how they moved. Voyager One didn't pick up many spots or other targetable features on Saturn, and Anne Bunker, who did most of the detailed work, and I had only a couple of weeks in June to track them using Voyager Two's pictures before we had to do our targeting. It's exciting to see our targets turn up now in the pictures. The trouble is, their arrival coincided with the meeting, so each time one appeared on the screen, I snuck out and made a print. We seem to have missed only one—the rest are all there. Apparently it's easier to predict Saturn's weather than the Earth's! Here's a picture of a white spot we call Anne's Spot—Anne found it. Here's Brown Spot One, an

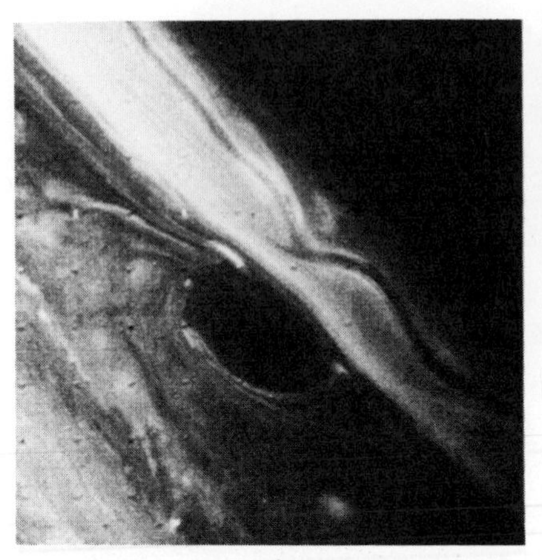
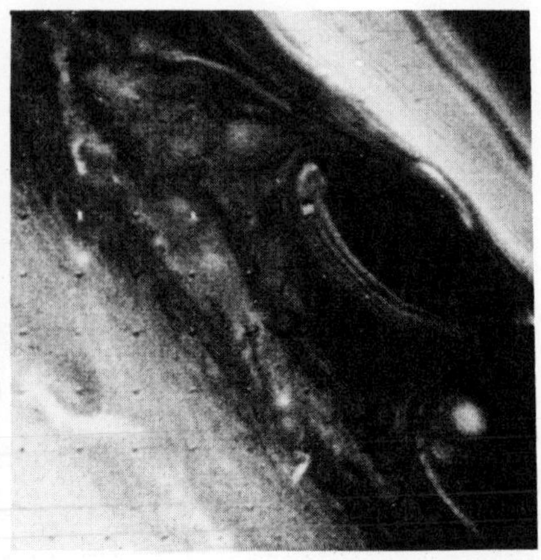

LEFT: *Brown Spot 1, imaged August 23, 1981.* RIGHT: *And imaged again August 24th, after an interval of ten and a quarter hours. From distances of 1.7 million miles and 1.4 million miles. Targeting had been done six weeks earlier.*

anticyclonic storm. Just north of it you can see what looks like a ribbon of material. That's a wind current, moving at a hundred and fifty meters a second—faster than three hundred miles an hour. What I'd really like to know is what the wind speeds are in the southern hemisphere—but we won't be able to tell that until Voyager Two is leaving Saturn, because the tilt of Saturn's rings is hiding the south from us as we approach. We have a pretty good idea about the pattern in the north, where alternating winds go all the way up to the seventy-fifth latitude, with the widest zone straddling the equator, but Voyager One gave us only a faint idea of the wind pattern in the south, and I want to establish whether there is a symmetry between north and south.

"Symmetries are good to look for in science, whether or not they might support a theory, and in this case there are, as you know, two conflicting theories about what drives the winds. If the internal theory is correct, that the Sun's heat is driving them—Verner Suomi and Garry Hunt, two of my colleagues in the planetary-atmosphere group, tend to think along these lines—we would expect to see some symmetry between the wind patterns in the northern and the southern hemispheres, though it wouldn't have to be very precise. One reason that I am not partial to the external theory is that only the top hundred miles—about as far down as the Sun's heat penetrates—would be involved; in other words, it deals with Saturn as though it were a planet with a thin atmosphere over a different sort of surface, like the Earth, and ignores what we know: that Saturn is a ball of gases with a depth at the equator of some thirty-seven thousand, three hundred miles, with heat sources of its own." Ingersoll had become interested in work done by a physicist at the University of California in Los Angeles, E. H. Busse, which suggested that fluid in a rotating planet, such as Saturn and also Jupiter, would form internally a series of coaxial—that is, concentric—cylinders, with the polar axis at their center. According to this idea, the alternating wind zones that circle Saturn latitudinally are, in effect, the exposed, rounded-off tops and bottoms of a number of cylinders, nested one inside the other; they would phase slowly from one to the next, without any abrupt transition (the way, indeed, the winds do), and they would spin at progressively faster rates, the innermost, halfway down, spinning the slowest. "With this model, we would expect to see quite exact symmetry north and south of the equator, with respect to location of the winds, their direction, and their speeds—which should get less as you move away from the equator," he said. "Jupiter—another gaseous planet—seems to fit the model, and Saturn seems to be shaping up that way, too. Because the curved surface of the planet would cut longitudinally *down* the outermost cylinder, instead of rounding off only its top and bottom, you would expect the wind zone at the equator to be the broadest—and indeed this is the case, both on Saturn and on Jupiter. Also, you would expect the alternating wind zones to go to higher latitudes on Saturn

Anne's Spot.

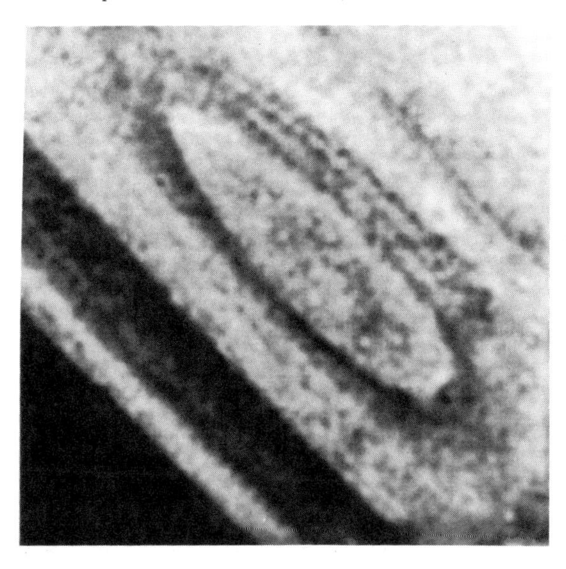

than on Jupiter, because Jupiter has a relatively larger metallic zone surrounding its core, and there would be no cylinders where the hydrogen becomes metallic. This also seems to be the case."

Monday morning, the day before the Saturn encounter, an enterprising salesman had set up a booth just inside JPL's main gate and was doing a very brisk business selling Voyager posters, Voyager T-shirts, and wooden Voyager plaques. The long line that had formed to buy these items included no imaging-team scientists, most of whom were in their offices, poring over pictures that might in the future make it onto a poster or a T-shirt. I found Terrile back in the same interactive room, where he had apparently spent most of the night looking for embedded moonlets. "No luck," he said. "Jeff Cuzzi went through the last two-thirds of the pictures, searching the Cassini division for moonlets as small as ten kilometers in diameter, and found nothing. I went through about one-third of the pictures, looking in both the Cassini division and the C ring, and didn't come up with any moonlets either."

Cuzzi had gone home, presumably to get some sleep. "He was getting worried yesterday afternoon about not finding anything," someone said.

"You should have seen him last night! He was practically climbing the wall," said Terrile, turning back to the interactive, where Hyperion,

Hyperion looked like a potato. LEFT: *Imaged August 23, 1981, first from 1.1 million miles.* RIGHT: *And later from seven hundred and forty thousand miles, so that the resolution increased: the smallest visible objects went from twenty miles to twelve miles across.*

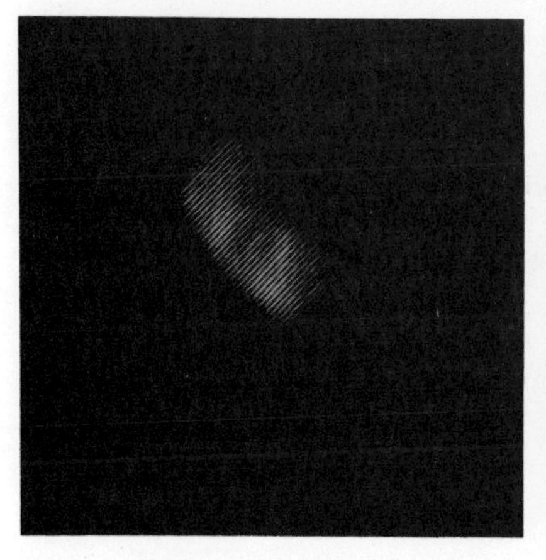

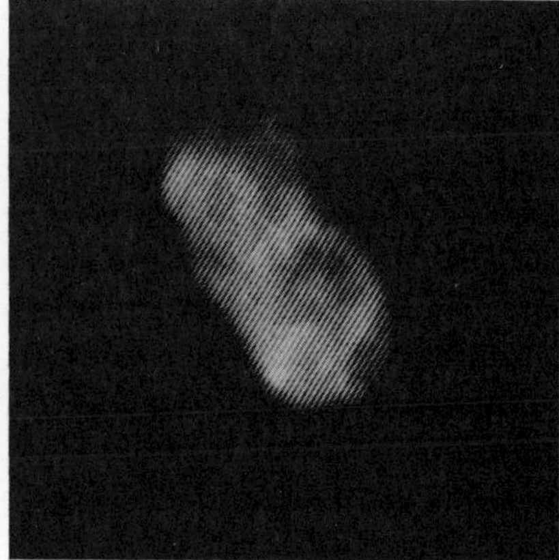

Voyager 2's next encounter, occupied about a quarter of the screen, looking rather like a potato. (Hyperion is irregularly shaped, like many asteroids and small satellites. A moon's shape seems to depend on its origin; if it accreted from material in the primordial planetary nebula, it would tend to be spherical, but if it is a fragment of a parent body that broke up, it would not necessarily become a sphere.) The encounter with Hyperion would occur that evening at 6:27. Torrence Johnson came in to have a look and said, "Well, Hyperion's turned out to be more interesting than we thought. Not only is it oblong—I'd call it bar-shaped—but it looks as if the long axis is pointing in the wrong direction. You'd expect it to be pointing at the center of Saturn, which would be the case if it was tidally locked to the planet, like most of the other moons, which are icy." A moon made entirely of ice, more apt to be slushy inside than one made primarily of rock, is easier to get into tidal lock because the planet's gravity causes the moon to dissipate its rotational energy by tidal friction: the slushy interior tends to drag on the rigid exterior. In such a case, the planet's gravity will also cause the moon to bulge, making it easier to lock onto it. Carl Sagan stuck his head in at the door and, noting a large crater on Hyperion, speculated that at some time in its past the satellite had been in tidal lock and that the impact had knocked it out of its rigid position.

"If it had been hit as recently as a few million years ago, it could still be oscillating," said Terrile. "It isn't tumbling, though, because in that case we would have seen a periodicity from Earth—an alternate brightening or darkening as the larger face and then the smaller face was toward us—but Dale Cruikshank, who's been studying Hyperion's light curve over time, has not had the chance to see a periodicity, if there is one. It'll be a while before we can tell much. We'll have to wait until pictures from different angles are in, to get an idea of its true shape and what axis it's spinning on."

At about seven o'clock that evening, when the first of ten pictures of the Hyperion encounter were arriving, only a smattering of scientists were in the interactive room. With the spacecraft some three hundred thousand miles from Hyperion, the encounter was not considered an especially momentous event. When Hyperion flickered into the middle of the interactive screen, Joseph Veverka, the small-satellite-and-asteroid expert from Cornell, said, "It's looking more the way we would expect it to." Hyperion appeared somewhat rounder than it had in earlier pictures, because Voyager 2 had now come up over a different side; as a result the moon had taken on a different shape. What had appeared to be a potato turned out to have been an edge-on view of a larger object

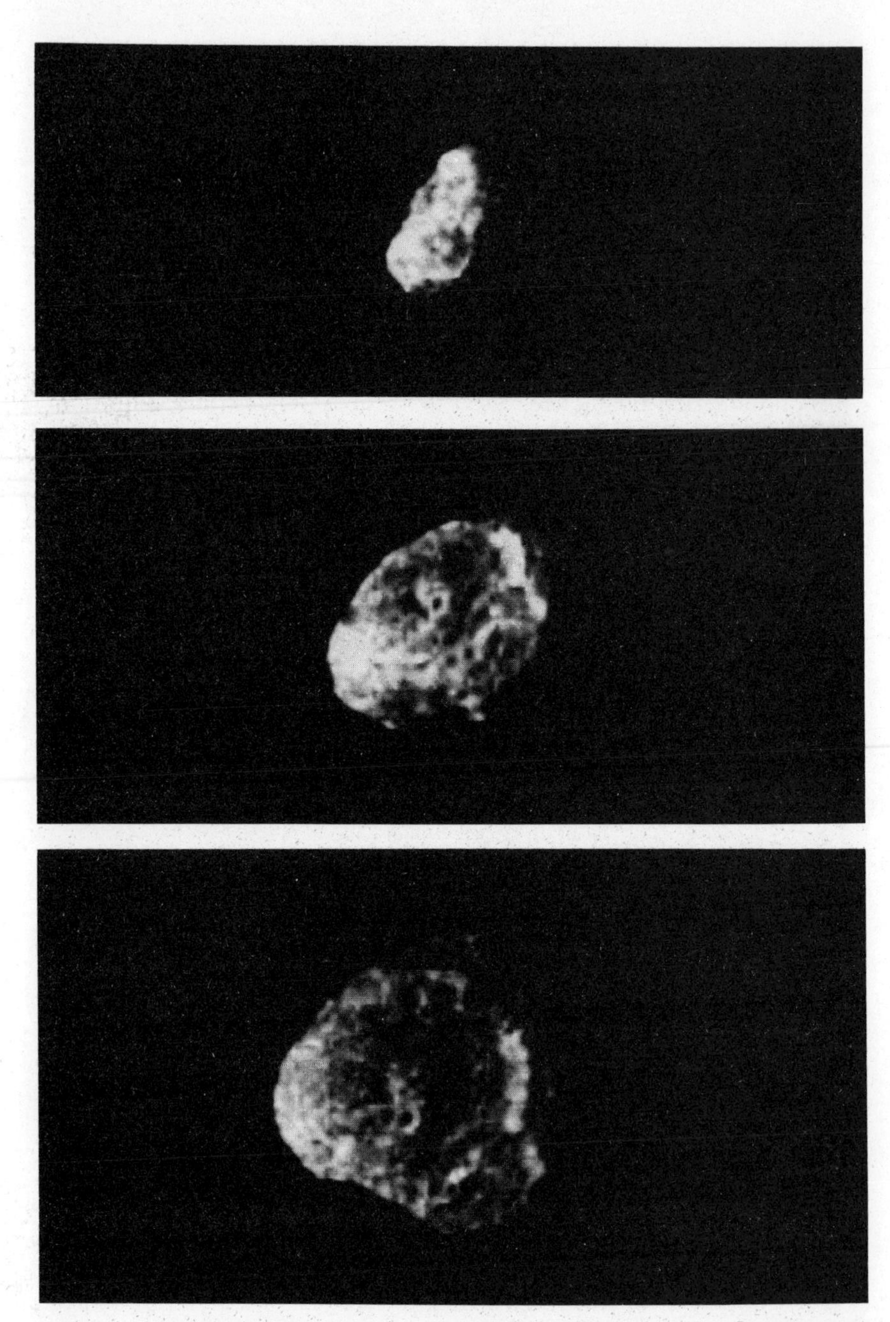

Hyperion: What had first appeared as a potato turned out to be a hockey puck—but which way is north? Imaged August 23, 1981, early morning of August 24th, and noon of August 24th, from seven hundred and forty thousand, four hundred and thirty thousand miles, and three hundred and ten thousand miles.

shaped something like a hockey puck; and as the figure (or, rather, the imaging team's perception of it) changed shape, the long axis changed, too. As far as Veverka could tell, the new axis didn't seem to point at Saturn either. "Hyperion is very irregular, and that suggests that it's very rigid," he said. "If it's not in tidal lock, that would indicate that it is primarily rock rather than ice or that if it *is* ice, then the ice inside isn't slushy." Hyperion had splotches of dark material randomly distributed on its surface—something that was of interest to the Phoebe-dust people, for if dust from Phoebe had spiraled on in past the orbit of Iapetus and settled on Hyperion, then, since that moon probably wasn't tidally locked, the dust might be expected to land randomly. (According to an expanded Phoebe theory, any black dust that escaped Hyperion would be swept up by Titan, the next moon in line, and would disappear into its atmosphere—which does in fact contain a number of carbon compounds—and this would explain why the inner moons were unsmudged with Phoebe dust.) There were others who felt that Hyperion's rotation was not sufficiently established to draw any conclusion as to whether or not it was tidally locked, and some who thought that the dark splotches there simply bore out the theory that only in the outer reaches of the primordial Saturnian nebula had it been cold enough for carbonaceous material to be incorporated into the body of a moon. (Over a year later, new evidence from Earth observations seemed to indicate that Hyperion was rotating lazily, at the rate of about once an Earth-year.)

About an hour after the encounter, I ran into Veverka on his way to dinner. He was, he said, glad that he had waited for the encounter. "Hyperion was the first satellite discovered outside Europe—the first major discovery by an American astronomer," he said. "George Bond, of Harvard, sighted it in 1848. It really got American astronomy started. Never mind that an Englishman discovered it at the same time."

Tuesday, August 25th, the day of the Saturn encounter, I arrived at the imaging team's quarters at about eight o'clock in the morning to find that the action had already started, even though the encounter with the planet was some twelve hours away. "There's unbelievable stuff coming in!" said Terrile, who was firmly entrenched at the keyboard of one of the interactives. "I feel guilty that I got some sleep last night. You missed the big excitement, which was the discovery of a huge crater on Tethys, like the huge one on Mimas. Tethys is around six hundred and fifty

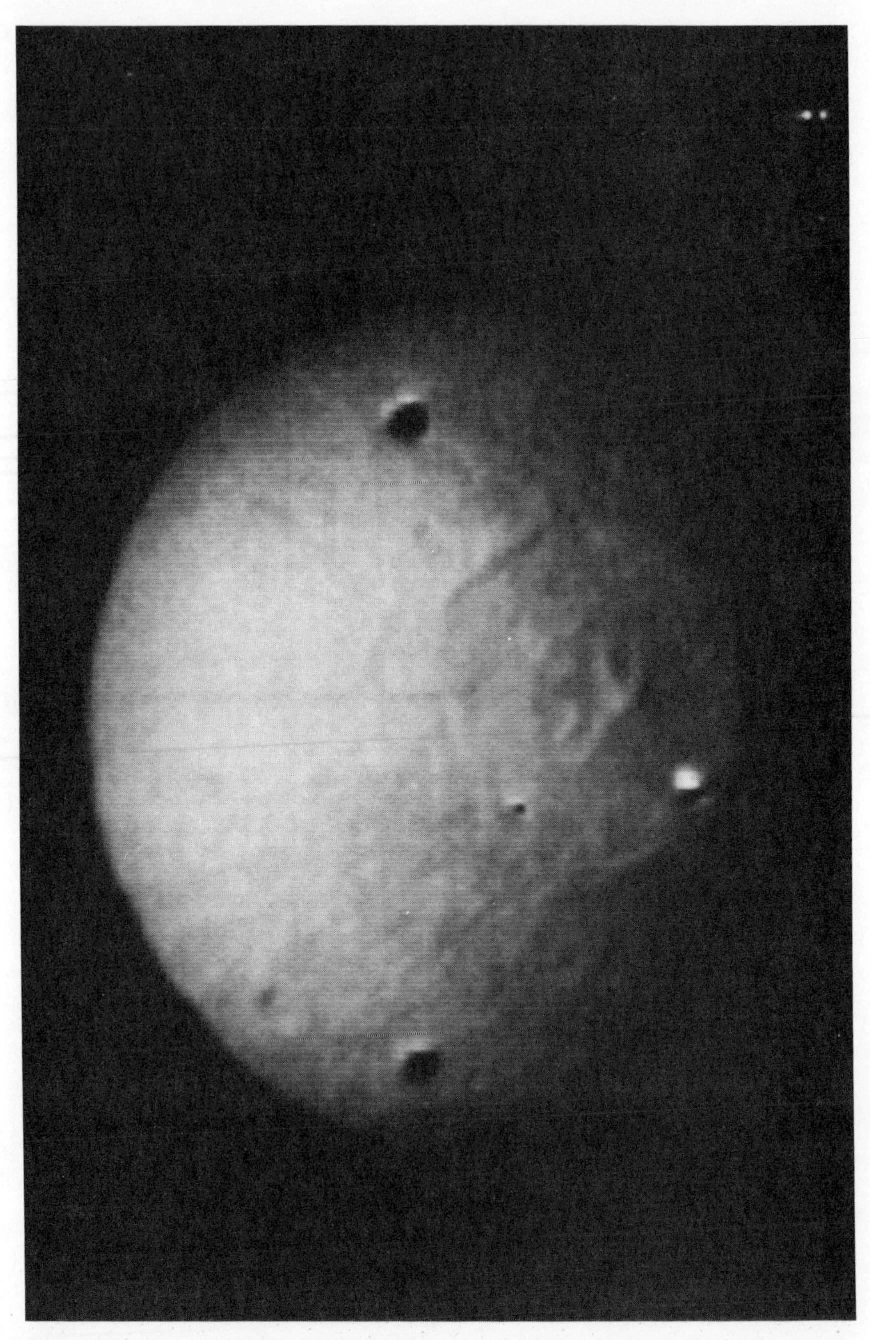

The huge crater on Tethys, as seen the day of its discovery, August 25, 1981, from a distance of six hundred and twenty thousand miles. The crater is two hundred and fifty miles in diameter.

The giant crater on Tethys, imaged August 25, 1981, but processed for contrast and relief by the Image Processing Laboratory and released several days later.

miles in diameter, and its crater is two hundred and fifty miles across. The whole crater is bigger than Mimas—the entire satellite. And the crater's diameter is even bigger in relation to the satellite than that of any other crater in the solar system, including the one on Mimas, which we had thought was made by the largest impact, relatively, that a moon could sustain without breaking up." (The fact that the two giant craters were so close to the size that would have shattered their moons was, Shoemaker told me later, further indication that many of the moons might have been shattered many times and reassembled, particularly the smaller, inner ones where Saturn's gravitational focusing was a factor. A further proof, he told me still later, emerged when the crater counts for both missions were complete, for it turned out that the bigger, outer moons, such as Iapetus, Dione, and Rhea, had a greater proportion of the larger, older Population I craters; whereas tiny Mimas and Enceladus, close to Saturn, were covered mainly with the more recent Population II craters—strongly suggesting to Shoemaker that those moons had lost their Population I craters as a result of being shattered.) The giant crater on Tethys, I found out later, did not coincide with the mysterious round feature seen during Voyager 1's approach to Tethys.

Reta Beebe, of the planetary-atmosphere group, stuck her head in the door and said, "Verner Suomi has made a discovery. He says he's got everything all solved."

I said I'd be right there; but, before I could go, Terrile had switched on a new picture of the rings. "Oh, there's so much wonderful stuff to look at!" he exclaimed. "Let's take a look at the big gap in the A ring; it's the second biggest gap in the rings, after the Cassini division." He flashed the picture on the screen and then enlarged it. "Look! There's a ringlet that Voyager One found," he said. "This is a much better view of it than Voyager One's." He peered closely at the ringlet. "Do you know, I think we've got another F ring here—this one really *is* kinky. We can certainly *use* another F ring, because the old one isn't living up to its billing. Maybe it just got up and moved into the A ring's gap. It's unbelievable! I'm afraid to go on. I don't know what will turn up next." He left the picture on the screen; the ringlet, a faint wisp etched against the black background, looked a little like a snarled fishing line.

"Wonderful stuff to look at": A three-thousand-seven-hundred-mile swath of the B ring, imaged the morning of August 25, 1981, from a distance of four hundred and sixty-one thousand miles, revealing the ring structure broken up into about ten times more ringlets than previously suspected. The narrowest features are about ten miles wide.

I found Verner Suomi in an office he shared with Andrew Ingersoll—down the corridor and around a corner from the two interactive rooms. Suomi, whose silvery hair partly obscures his broad pink face, was talking excitedly with Garry Hunt, and they were both looking extraordinarily pleased with themselves. It turned out that both had been there since four o'clock that morning and were now running on a combination of nervous energy and exhilaration. I asked them what they had discovered.

"You tell him," Suomi said to Hunt. Despite the fact that Suomi has been involved with the study of planetary atmospheres probably longer than anyone else and has designed a number of instruments used aboard spacecraft, he looked as if he were a graduate student who had made a discovery that had eluded the professionals, and was about to burst with excitement.

"At about five o'clock this morning, we were looking at some brown spots—they rotate, like eddies—moving in one of the wind zones," Hunt said. "As they moved down the wind current, they began to come apart, that is, each spot became a *pair* of spots, traveling side by side but rotating in opposite directions. Now why does that happen?"

The kinky, elliptical ring in the big A-ring gap, imaged on August 25, 1981, in two different places, to demonstrate its eccentricity in relation to the otherwise circular A ring. The image has been much enhanced to bring out the ringlet, but at the sacrifice of other structures within the A ring.

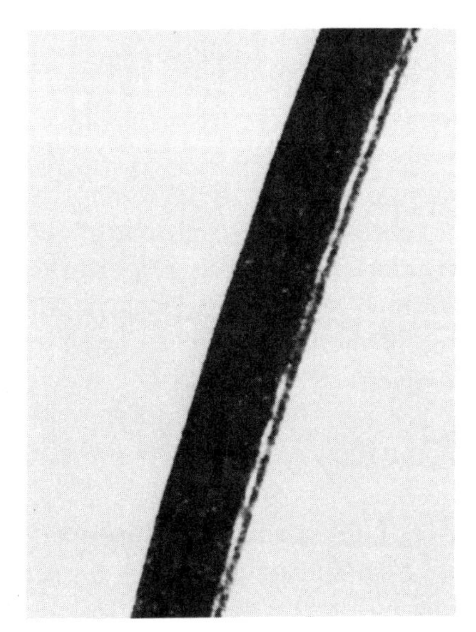

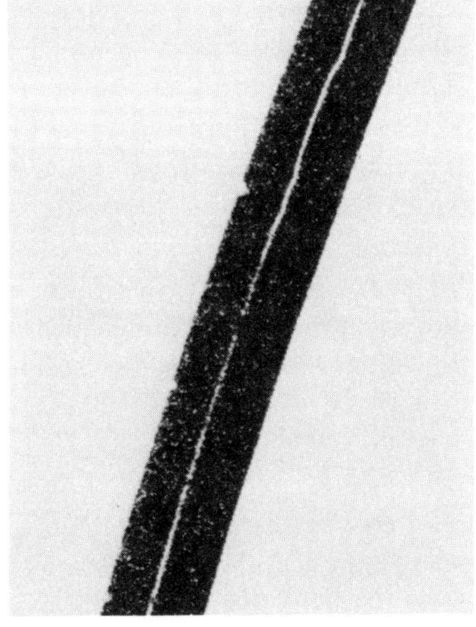

"Because other spots, which move at a slower rate—the white ones, which are convection currents welling up from inside Saturn—form barriers to the flow of the prevailing wind!" said Suomi, unable to contain himself. The convection currents, he told me, are the tops of vertical chimneys of warmer gases moving up from inside the planet, rather like upwellings in a pot of boiling water, and they apparently act like rocks in a stream. The eddies form pairs downstream of a convection current (or *column*, really) making an ever-broadening V as they move apart from each other and spin away. (The V's, Reta Beebe told me later, were the same features she and Ingersoll, at the time of Voyager 1, had called chevrons.) Each pair, rotating in opposite directions, and faster than the prevailing winds, move farther apart as they extend the V, and finally settle at the borders of the wind zones to the north and south, imparting their rotational energy to them in the manner of rollers propelling conveyor belts. (Later, when I discussed all this with Andrew Ingersoll, he told me that he thought the rocks-in-the-stream observation was interesting, but did not get to the heart of what drove the winds and the alternative zones. Clearly he still preferred his coaxial-cylinder theory and he thought the theory might account for the little white convection currents, too. They might originate from the cylinder underneath the cylinder of the wind zone they were in; thus you would expect them to move at a slower rate than the prevailing winds.)

Outside in the corridor, Joseph Veverka was mounting on a large piece of white cardboard several pictures taken of Hyperion from different distances and perspectives—a series of eight in all, each image bigger than the last as the spacecraft got closer, and then progressively smaller as it passed on its way. "We're still working on the rotation of Hyperion, to determine whether or not its long axis points to the planet," Veverka said. "Right now, one of the theories is that most of the time Hyperion *is* locked on Saturn, but that its orientation shifts whenever it passes Titan, which it was doing at the time of our close approach. (The following January, when the imaging team published its report of the Voyager 2 mission in *Science*, the most widely read professional scientific journal in the United States, the matter of whether or not Hyperion was tidally locked to Saturn was still unresolved.)

When I returned to the team quarters I found that John Spencer—who (along with Nicholas Schneider) had once again persuaded his professor, Robert Strom, that he would be indispensable to the mission—

had taken over the controls of one of the interactives. He was surrounded by a large group of scientists, including Soderblom, Owen, Terrile, and Johnson, as well as Strom, who was doing most of the comparative crater counts and measurements for Voyager. They were waiting for a picture of Enceladus to come down; although Voyager 2 was by no means as close to Enceladus as it would get, it was some 285,000 kilometers away, or about as close as Voyager 1 had been when it took its closest pictures, which had had a resolution of eleven kilometers; presumably Voyager 2's cameras, with their superior optical system, could do better. Enceladus had attracted attention as far back as Voyager 1's encounter with Jupiter, in March, 1979, when Io, one of Jupiter's moons, was found to be so hot inside that active volcanoes were shooting plumes of material from its interior a hundred miles or so into space. Io proved to be the most active planetary body in the solar system. Its heat is caused by tidal effects. Io's orbit is elliptical, because it is in resonance with a larger Jovian moon, Europa; the changes in gravitational stress as Io moves alternately nearer to and farther from Jupiter cause friction, presumably by slippage along faults, or cracks, in its surface, producing heat. Not long after Voyager 1's encounter with Jupiter, Charles Yoder, a JPL scientist who specializes in celestial mechanics, predicted that the same processes might be at work on Enceladus, whose orbit around Saturn was elliptical because it was in resonance with the larger moon Dione. Though no one was predicting active volcanoes for Enceladus (the forces involved were much less than at Io, because Enceladus is smaller, its orbit is less elliptical, and Saturn is a less massive planet than Jupiter), evidence of some heating was visible, since Voyager 1, which could have resolved craters ten kilometers across, in fact revealed none—arguing for a moon that was relatively too warm and plastic to support them. Enceladus's greater reflectivity, which made it look brighter than the other moons, also seemed encouraging—though no one quite knew why.

After Voyager 1's Enceladus pictures had come in, Richard Terrile and Allan Cook, an astronomer from the Smithsonian Astrophysical Observatory, had gone on to make another suggestion, based on Yoder's, in an attempt to explain another mystifying phenomenon: the apparent association of Enceladus with the broad, hazy E ring, which extends from just outside the orbit of Mimas certainly to the orbit of Tethys and very likely to that of Rhea; Enceladus orbits suspiciously inside its thickest part. Very possibly, Terrile and Cook argued, Enceladus's smooth crust was very thin, and in that case any sizable impact would puncture a hole in it, releasing water vapor to space, which would freeze into the very small particles that make up the E ring.

"From the infrared measurements, Cook and I had speculated that the E-ring particles were extremely small—a few microns at the most," Terrile had told me at the time of Voyager 1's encounter. "That has an important implication. It means they must be very young, because micron-size particles in orbit around a planet don't survive long—they tend to migrate inward. Hence, they have to be replenished, and this requirement fits in very nicely with the theory that the E ring comes from Enceladus. In the pictures from Voyager One, it looks as if something were constantly modifying the surface of Enceladus, so that its craters are constantly being erased. One impact puncturing Enceladus's crust every few thousand years would throw enough water into space to maintain the E ring. And if some of the water fell back onto Enceladus, that would explain why the moon is so white, as well as so smooth. I must say I was happy when no surface details showed up in the pictures, for it fits in with what we've been thinking. Before the Voyager One encounter, Ed Stone, the chief scientist, asked some of us to talk to an artist who was drawing views of what the surfaces of the different moons might look like. When he asked about Enceladus, one scientist advised him to put a large crater in the picture. I told him not to—I told him to draw a flat, featureless landscape, no craters, no lineaments, no *anything*, and I hope he took my advice! He should be making a nice little reputation as a prophet."

The artist would have been as well off if he hadn't taken that advice, it turned out. When the picture that the scientists were waiting for appeared as a raw on a television monitor overhead, Spencer quickly called it up on the interactive so that it could be enhanced. At first, Enceladus appeared as a bright, smooth object—still featureless, and as fuzzy as a tennis ball. The picture had been taken from a distance about 20 percent farther than the Earth is from our own Moon; Enceladus's diameter—three hundred and ten miles—is only about a seventh that of our Moon. Spencer pressed some more buttons, and the enhanced picture popped onto the screen, revealing a much changed moon; Voyager 2 was observing a different side from Voyager 1.

"There's what looks like a series of ridges," he said. "And there are lots of craters—here, here, and here." Besides the ridges and the craters, though, there were broad areas where the ground was flat, smooth, and featureless. "The surface of Enceladus is the most variegated we've seen," Spencer went on. "The variety confirms the idea that something dynamic is going on here—in places, something has apparently been erasing features. But why speculate now? By this evening, we'll have pictures far better then these."

"That's never stopped anyone from speculating before," Strom said, peering closely at the screen.

I asked Spencer if any Io-like plumes shooting from Enceladus would be visible at this distance; he said that they would, and that there weren't. When I asked Terrile how he thought his and Cook's theory looked now, he said, "Some parts of Enceladus's surface have lots of what seem to be very old craters, and that looks bad for our theory since it suggests inactivity, but other areas are entirely smoothed over and young, and that looks good for our theory. Maybe we can remodel it a bit. Maybe we don't have to have a body that is entirely resurfaced to account for the E ring—maybe one that's only partly resurfaced would do it. I have to think."

Robert Strom was still bent over the interactive screen, studying a picture of Enceladus, when his name was paged over the public-address system. He had a phone call. His teammates laughed; more Enceladus pictures were coming in. "He's at lunch!" Strom shouted, and huddled closer to the screen.

In the corridor, I met Eugene Shoemaker and asked him what he thought about Terrile and Cook's theory, and also whether he could tell how warm Enceladus might be. If anyone could gauge a moon's inner warmth from its outward appearance, it was Shoemaker, who had said earlier that Tethys was once warmer than Mimas. He had deduced this from the fact that the floor of the giant crater on Mimas was still concave from the impact that had formed it, whereas the floor of the newly discovered giant crater on Tethys had resumed the curvature of that satellite, arching tens of kilometers above the crater's rim—an indication that Tethys had been warm and plastic enough to rebound. To my question about Enceladus, Shoemaker replied, "The question is: If Enceladus really has been undergoing the large tidal heating we think it has, then what would be the fate of any craters on its crust? That would depend on the thermal gradient of the body—how much heat is in it, and how fast the heat works its way to the surface—and how long it would take the craters to flatten away as a result of the plastic flow of the crust, which of course is correlated with the warmth. The rate at which the craters disappear depends on their size, too—the bigger the crater, the sooner it will flow away, since the gravity flattening it down will exert more force. Therefore, the rate also depends on the gravity at the surface of Enceladus and on the viscosity—by which planetary geologists mean the rigidity—

of the crust, which in turn is dependent not only on the amount of internal heat but also on the materials the crust is made of. We have a formula that accompanies all these factors. Viscosity is a function of temperature. Since it gets hotter as you go deeper, viscosity will decrease (the plasticity will increase). In other words, you could have a rigid outer shell, or crust, over a warm, soft interior. The crust will creep, and craters will disappear in anywhere between ten thousand years and billions of years, depending on what all those numbers are that go into the formula. If I knew the amount of tidal energy going into Enceladus, I could make some pretty good predictions about the viscosity of its surface. I'd also need to know Enceladus's structure. If it is a rigid body, the tidal energy wouldn't create much heat, because there wouldn't be much slippage and friction, but if the inside is squashy, there would be many more cracks, thus more friction, thus more heating. That's why I'm excited about the high-resolution pictures we'll get tonight: after I do a crater count, knowing the cratering rate around Saturn, I can deduce the degree of tidal heating; once I have that, I can make a pretty good guess about Enceladus's internal structure. So I'm waiting with bated breath; the pictures will be coming in just before Voyager slips behind the planet."

In the other interactive room a little later—about eight hours before that evening's close encounter with Saturn—Terrile, Cuzzi, and a young associate of Cuzzi's, Jack Lissauer, a slight, dark-haired man who was a graduate student in astronomy at the University of California at Berkeley, were looking at a set of four pictures of Saturn's rings, taken from a range of 388,375 miles, which had begun to arrive at 11:45 A.M. Cuzzi made a hard copy of part of the A ring, just outside the Cassini division, which he said had some wonderful examples of what he called density waves, a phenomenon, by no means clearly understood, that had been predicted for the rings by two astronomers, C. C. Lin, of the Massachusetts Institute of Technology, and Frank H. Shu, at Berkeley (where he is Lissauer's professor—previously, Shu had been at Stony Brook, where he had taught Terrile). Goldreich was also involved. According to Lissauer—one of whose reasons for being present was the density waves—they are a delicate series of ripples, apparently going all around the rings, diminishing as they move outward, very likely from a resonance orbit. The resonance would provide a sort of periodic forcing that would get the rippling started; the gravity of the ring particles themselves would

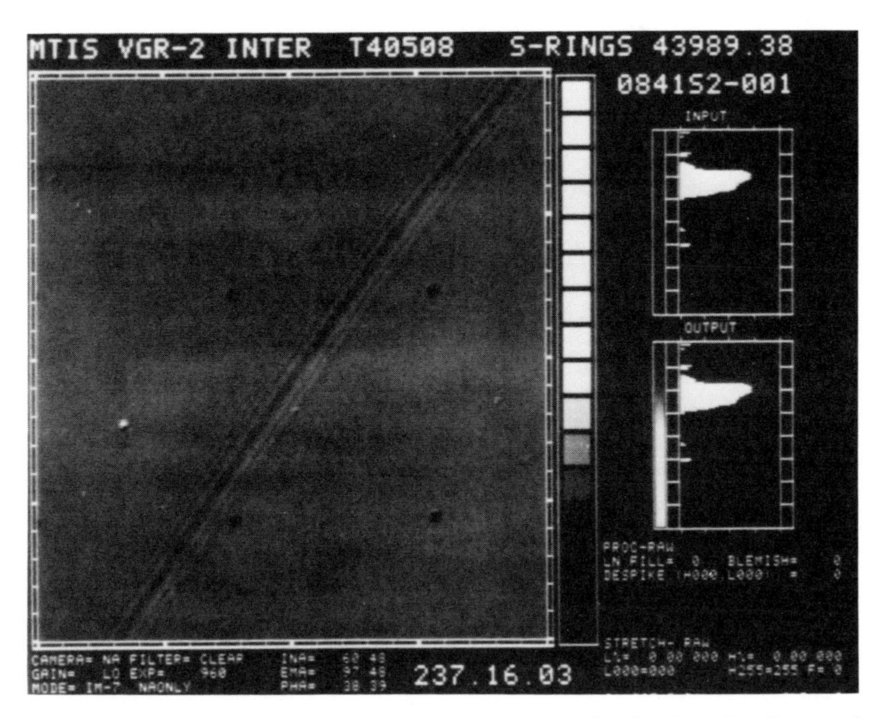

An interactive picture, from imagery taken August 24, 1981, showing what Cuzzi and Lissauer think are density waves moving outward from a resonance gap. Ripples moving inward from the gap may be another type of wave.

provide a counterforce pulling the ripples back again. According to Lin and Shu's theory, a similiar sort of phenomenon had been observed in galaxies, whose billions of stars often form spiral, rippling patterns—indeed, one of the things he would be looking for, Lissauer said, was whether the density waves in the rings spiraled, too. (They do; Cuzzi would later liken them to a watch spring.) Part of the fascination of the rings, of course, is that they represent a primordial and basic form of matter in space—the disc of particles, whether on a galactic, stellar, or planetary scale, is a phenomenon present all over the universe—and Saturn's rings were the example closest to home, where the phenomenon's structure, dynamics, and behavior could best be observed.

Cuzzi said he was glad these particular density waves had turned up, because the only other definite set of them—a series of ripples seen by Voyager 1 near an Iapetus resonance in the Cassini division—had not been seen yet. Whether the ones that were now visible, at a Mimas resonance in the A ring, were new, or whether the inferior camera on

Of the three prominent sets of ripples—all of them on the A ring, inward from its big gap—the outermost are density waves associated with the 5:3 resonance with Mimas, and the innermost are density waves in resonance with one of the shepherding satellites. The dark set in the middle are a different type of phenomenon called bending waves, not identified until after the Voyager 2 flyby of Saturn.

Voyager 1 had missed them, he couldn't say. It was too early to tell if there were any other changes in the main rings—a possibility Terrile had been looking forward to investigating ever since Voyager 1 left Saturn nine months ago.

Before the next batch of ring pictures came down, there was a brief interlude when a couple of what were called "search photographs" arrived. They were reconnaissance photos to look for some possible new moonlets near Tethys—a promising place to ferret around in because two new moonlets had already been found, telescopically from the Earth, by Bradford Smith and some associates at the observatory of the University of Arizona at Tucson, in 1980, when the Earth itself was passing through Saturn's ring plane (which it does about every fifteen years), and hence the rings, seen edge-on, reflected no light—the same circumstances that had existed in 1966, when the E ring had been discovered. One was sixty degrees ahead of Tethys on its orbit, and the other sixty degrees behind; they were at two of Tethys's Lagrange points, places which (for reasons having to do with the interaction of the gravity fields of a satellite and a planet) attract matter into curiously shaped little orbits. Presumably, in the early bombardment that had resulted in the giant crater, Tethys had had chipped from it large chunks of ice, which ultimately would have fallen back onto the satellite—except for any trapped in the Lagrange points. Dione, of course, had a small satellite in one of its points, Dione B, and of course Voyager 2 was imaging similar spots near other satellites as well, though so far without luck. There was reason to think that there might be more than one satellite in Tethys's trailing point—which the scientists were looking at now—and indeed as they peered into the screen, they saw three dots where there should only have been one. On the next picture, of the same area, there were *four*.

"They've multiplied," said Owen, who had wandered in from the room next door—there was much movement back and forth between the two interactive rooms.

Hard copies of the two pictures were made, and, upon comparison, two of the four dots were immediately pronounced blems; one of them, however, was thought to be a newly discovered moonlet. It wasn't; later, when I spoke with Candice Hansen, the team's expert on targeting satellites, she said that the third spot had turned out to be a blem, too. (The absence of moonlets in this place was afterward confirmed.) When I expressed disappointment that I had not after all been present at the discovery of a new celestial body, she said, "Now you understand how *we* feel a great deal of the time."

Though members of the imaging team would, of course, continue to ransack the pictures visually for any new discovery, a more methodical search for the new moonlets in this and other pictures would shortly be undertaken by a colleague of Terrile's at JPL, Stephen Synnott, who had found a couple of new moons around Jupiter by going over the bits of data that encoded the brightness level of each of the pixels—analogous to the dots that compose a newspaper photograph—looking for small, somewhat brighter, areas that escape the eye. He had tried doing this with the Voyager 1 data, but without any luck—though there had been a flurry of interest in a couple of streaks near Mimas observed by some members of the imaging team, but which Synnott had subsequently ruled out. He was more hopeful of success with the imaging data from Voyager 2's better cameras. The results of his labors, though, would be many months in coming, as there are two hundred and fifty-six possible shades of brightness for each pixel, and for each picture there are six hundred and forty thousand pixels conveyed in five million data bits.

"This is going to be a great disc pack; everything's going to be on it," Cuzzi said, when the last satellite-search picture was turned off and the scientists were waiting to see what was coming next. "Now let's see what's of interest here."

A new series of thirty-five ring pictures was just beginning to arrive. The first picture in itself was enough to excite Terrile, for it included the F ring. It was still a single unkinky strand. "It would be nice if we got just one photograph of the F ring showing more than one strand," Terrile said, pressing a button to call up the next picture. "Good!" he said. "Look at that! It looks as if the F ring goes from one strand to two."

(About three months later, Terrile discovered that at precisely the point where that section of the F ring bifurcated, it was also braided—or, more precisely, was becoming unbraided—and he also found evi-

The F ring and its inner shepherding satellite, imaged August 25, 1981. The F ring is still a single, unkinked strand.

dence, in the form of streaks on some long exposures, that the F ring contained a number of tiny moonlets, about ten miles across, which, along with the shepherding moons, might cause the intermittent braiding and kinkiness; the micron-sized particles composing the ring would be easy to deflect. Presumably the same situation pertains to the kinky ring in the big A-ring gap, which Terrile later found was really *two* kinky rings; sometimes one was visible, sometimes the other, sometimes they both vanished. The other variety of elliptical rings—the smooth, unbraided, unkinked sort, of which there are four—Terrile thinks can be explained by just having the large particles, or small moonlets, inside them, but no external shepherds like the F ring that would excite the odd perturbations. Even though no bends had turned up at the point in the F ring where the two shepherds had most recently passed, such a conjunction still appeared to be the most logical way of kinking a ring. Further work strongly suggested that the kink might not occur until several orbits after the disturbance, just as a person who is pushed might not fall down right away; indeed, a kink is very likely caused by one shepherd in closer-than-usual proximity to the ring, while a braid is caused by a conjunction of two.)

The imaging-team meeting that afternoon was well attended. Bradford Smith announced that the new administrator of NASA, James M. Beggs, would be visiting the team's quarters in about half an hour to see some of the new pictures and it would be a good idea for most of the team members to take the opportunity to absent themselves, perhaps by attending the joint science-team meeting upstairs.

"What's the matter?" said Merton Davies, the geodesist from the Rand Corporation who was responsible for measuring the exact size and shape of the various satellites. "Doesn't he want to see science being done?"

"Yes," Smith said. "But we don't want twenty-five guys in the interactive room when he's looking at the pictures."

There was general grumbling, but everyone saw the point. The future of much of NASA's planetary-exploration program was still very much in doubt; the Venus orbital mission, which would attempt to make a radar map of the surface of Venus through its thick clouds and which a number of people present wanted very badly, was thought to be in danger, despite President Carter's last-minute support, because the new President had been elected on a platform to reduce taxes and cut government spending. Another mission—one that Smith, among others, was particularly concerned about—to fly alongside Halley's Comet, was in similar jeopardy. Indeed, the only mission definitely approved after Voyager was Galileo, which would orbit Jupiter to swing repeatedly by its moons and to drop a probe into the planet itself; though Galileo was tentatively scheduled for launching from the space shuttle in 1985, that date was not as definite as everyone would like it to be, particularly Torrence Johnson, who was to be its project scientist. "When we had some officials of the Office of Management and Budget here a couple of weeks ago, they practically had to be dragged away from the interactives," one scientist noted, clearly hoping that Beggs would be similarly impressed.

After Beggs had made his visit, he and other NASA officials held a press conference in Von Karman Auditorium, at which they talked with somewhat guarded enthusiasm about the planetary program. Beggs, a businesslike man wearing a light-gray suit, read a few remarks from a prepared statement. "I cannot conceive of NASA without a strong scientific program exploring the unknown," he said. "There is still a substantial amount of money for exploration—though not as much as a decade or so ago." He seemed unable, or unwilling, to put dates on any future missions. During the question period, when he was asked about funding for a crucial element of the Galileo mission—a Centaur rocket, which was to propel the spacecraft to Jupiter once the shuttle had put it in Earth orbit—all he said was, "I think the Reagan administration is generally sympathetic."

On the way out of the auditorium, David Morrison, who was then chairman of the planetary-sciences division of the American Astronomical Society, said, "The Reagan administration people keep talking about how much they like space, but they keep on cutting our budget. I wish they weren't so sympathetic: it makes it harder to argue with them."

That evening, in quick succession, there would be a number of encounters: at 6:05 with Dione, at a distance of 312,000 miles; at 7:34 with Mimas, at 193,000 miles; and at 8:24 with Saturn itself, at 63,000 miles. The approaches, however, did not arouse much excitement among members of the imaging team, for during the approach period (and including the Dione encounter itself), Voyager's scan platform—the movable tray, attached to a boom extending from the body of the spacecraft, that carried the cameras and three other instruments (all of which had to be aimed jointly)—would be pointed at the rings. This was being done to allow one of the instruments, the photopolarimeter, to track the light of two stars, Delta Scorpii and Beta Taurii—through the rings, which would give highly detailed information about ring structure. The annoyance of many members of the imaging team at the fact that much of the most scenic part of the mission would be occupied in this fashion was, of course, balanced by their own extreme interest in the ring structure; indeed, Edward Stone, the project scientist for Voyager, had ruled in favor of the photopolarimeter observation by following a principle he had established for such cases: always to choose the alternative that would provide the most new science. Dione and Mimas had been imaged at much closer range by Voyager 1, and although different faces of both would be presented to Voyager 2, Stone felt that quite a lot was already known about these moons, and as for Saturn, there were an enormous number of pictures of the planet from both Voyagers. Immediately following the photopolarimeter observations, the scan platform would swivel so that the cameras would be on target for Voyager 2's approach to Enceladus, which it would encounter at 8:45—a very close encounter as the spacecraft would come within fifty-four thousand miles of it—and to Tethys, which it would encounter at 11:12, at a distance of fifty-eight thousand miles. Pictures would be taken for mosaic maps of those satellites—seven of the Enceladus approach, and six at encounter. At 9:18, in between the Enceladus and Tethys encounters, Voyager 2—out just beyond the G ring—would pass beneath the ring plane, taking as it did so an edge-on picture of the rings that might reveal their thickness and might also confirm the theory that the spokes were small dust particles levitated above the B ring. (The possibility of catching a spoke edge-on was enough to throw Richard Terrile into a state of ecstasy.) A number of these pictures—the Enceladus encounter, the ring crossing, the Tethys encounter—would not be transmitted to JPL that night; rather, they would be stored on a tape recorder in the spacecraft and sent back the following morning. This was necessary because at 9 P.M. Voyager 2 would disappear behind Saturn and would not emerge until an hour and thirty-five minutes later; transmission would be tricky for well over an hour on either side of that period, because of interference created by the rings as well as by strong plasma and magnetic fields surrounding the planet. During part of this period—from 9:08 until 10:45—the spacecraft would be traveling in Saturn's shadow; pictures would be taken at this time of the dark side of the planet, which might reveal flashes of lightning, or even auroras, in Saturn's atmosphere.

I arrived at the team quarters that evening at eight-thirty, a few minutes before the first Enceladus pictures and just in time for the arrival of the two approach pictures of Tethys. (The latter had, of course, been taken almost an hour and a half before—at a distance of about a hundred and seventy-seven thousand miles.) Torrence Johnson and Eugene Shoemaker began studying them at once. Curling around one limb of Tethys was what looked like the tip of an octopus tentacle. It was one end of the gigantic trench that Voyager 1 had revealed on the far side of that satellite. The trench, which is about sixty miles wide, two and a half miles deep, and seven hundred and fifty miles long, is larger than any other canyon in the solar system except the Valles Mariner on Mars. (As the nomenclature committee for the outer solar system had decided on using names from the *Odyssey* for Tethys, it had decided to call the trench "Ithaca Chasma," and would recommend naming the giant crater found on Tethys by Voyager 2 for Odysseus himself.) Shoemaker, who at the time of Voyager 1's flyby had suggested that big impacts on one side of a moon might cause cracks on the opposite side—one example, Shoemaker thought, was the giant crater on Mimas (it would be dubbed "Arthur," the committee having picked that legend for that satellite), which was antipodal to rough, cracked terrain—wondered now whether the trench on Tethys might have been caused by the impact that had made the giant crater. (Though the two did not prove to be antipodal, later measurements showed that the trench bore a complicated relationship to the crater that suggested that there might well have been a common origin.) There were a couple of other explanations for the trench. Harold Masursky believed that convection inside Tethys might have pulled the crust apart, and Laurence Soderblom believed that, since Tethys had probably frozen from the outside in, the frozen crust had been split open by the expansion of the core as it froze; freezing in this fashion would increase the surface area by 7 percent—roughly the percentage of Tethys taken up by the trench.

At length came the awaited series of seven Enceladus approach pictures. The series moved up along the lit part of the moon toward its day-night line, which showed up from one picture to the next as an increasingly large crescent of black; they had been taken from seventy-five thousand miles away—almost three times as close as the pictures taken that morning and only twenty thousand miles farther away than the encounter, an hour and a half later, the pictures of which would be relayed to the Earth next morning. The ones coming in now would be almost as good; they would have a resolution of 1.9 kilometers—that is,

The trench on Tethys, curling around the limb like an octopus tentacle. Imaged August 26, 1981, but processed later to bring out surface detail.

they would pick up any features at least that size—compared with the 1.6-kilometer resolution expected later.

"Are these things trenches or mountains?" Johnson asked, pointing to some odd squiggles and bumps. They turned out to be both—channels with mountains on either side. The channels were quite sinuous—some of the scientists were reminded of grooves on Jupiter's giant moon, Ganymede, the biggest satellite in the solar system (it is a little larger than Saturn's Titan), which had never been satisfactorily explained. Moreover, only Ganymede had the variety of terrain exhibited by tiny Enceladus, one thirteen-hundredth Ganymede's volume. As Enceladus should have lost its heat long ago, all this variety, of course, confirmed the belief of Shoemaker and others that the moon had some sort of continuing heat source—most likely tidal energy.

"Enceladus is a weird moon," said Johnson, pointing out several different types of terrain (there were six, by later count). "It seems to me that Enceladus has taken one feature from each of several different bodies and thrown them together." One part of the moon was cut by three huge parallel grooves and a fourth at right angles to them, forming, appropriately, a gigantic E—a monogram, one scientist ventured, though he didn't know whether it stood for Enceladus or E ring. (Later, when I asked Masursky, a member of the nomenclature committee, what myth would be used as a source for names for the craters and other odd features on Enceladus, he said, "We'll need something special. For Iapetus and Hyperion, we'll be looking at *Beowulf*, the *Song of Roland*, and the *Tale of Genji*, which we didn't get around to using last time, but for Enceladus I've just sent for the sixteen-volume translation of the *Arabian Nights* by Richard Burton, which has an enormous glossary of names—Enceladus has an enormous number of features—but more importantly, the *Arabian Nights* is a mysterious book, which is suitable for such an exotic satellite." Eventually, the committee indeed recommended the *Arabian Nights* for Enceladus, for the reasons suggested by Masursky; the *Song of Roland* was recommended for Iapetus; and Hyperion was

Enceladus, as it was seen the evening of August 25, 1981, not long before the encounter—"a weird moon," complete with its own monogram.

slated for a worldwide assortment of gods and goddesses of the Sun and the Moon.)

Some craters adjacent to smooth plains on Enceladus appeared to have been cut in half where they met the plains; other craters appeared to have been pulled out of shape, so that they resembled smoke rings in a breeze. Shoemaker took a particular interest in these, of course. When I asked him for a little further analysis, he brushed his hands over a hard copy of one of the pictures, as if to feel the various types of terrain, and pointed to an area containing a great many craters, some of the larger ones quite distorted. "A terrain with that many craters would have to be at least three billion years old," he said. "It would take that amount of time, too, for the large craters to become wobbly. I'd say you've got an icy crust about twenty kilometers deep, and you could ding that—puncture it—with a good impact, which means that Enceladus could indeed be the source of the E ring. Underneath the crust, it could well be molten. These are just my first impressions, though—I'll probably revise them later."

Soon the raws stopped coming in. The spacecraft was about to be occulted from the Earth, and most of the scientists went home to bed, in order to be fresh for the transmission of the recorded data next morning.

OUTBOUND AGAIN

When I arrived at the imaging team's quarters Wednesday morning, there were more scientists, and less noise, than usual; the members of the imaging team stood quietly looking at the monitors, amid the unaccustomed hush. The door to one of the interactive rooms was shut, with a sign on it that read "Keep Out."

I found Richard Terrile in the other interactive room. "Are you aware of what happened last night?" he asked. "Around midnight, some of us went over to the pressroom in Von Karman Auditorium for champagne, to celebrate the acquisition of signal from the spacecraft, which came in at one minute after twelve. Half an hour later, when the raws started coming down, somebody noticed they were blank. The scan platform was locked at a peculiar angle in azimuth—that's its lateral, circular axis as opposed to its axis of elevation—so that the cameras and the other instruments on it could not be aimed in that rotation. They were looking at space! We haven't had any pictures since." The scan platform is attached by a rod at its center to the end of a boom extending from Voyager's side. On top and to one side of the rod are the cameras; on the other side is the ultraviolet spectrometer; below the cameras is the photopolarimeter; and beneath the ultraviolet spectrometer is the infrared spectrometer. A motor and gearbox on the boom swivel the rod so that the platform can be rotated in azimuth; another motor and gearbox tilt it up and down to aim the instruments in elevation.

Terrile told me that when the blank raws began coming in, he and Cuzzi and Soderblom were among the few imaging-team members still around; most of the others had gone home. Even the normally indefatigable John Spencer and Nicholas Schneider, who had held the record for staying up at the time of Voyager 1's encounter, had left before the acquisition of signal. "Things go so well you get confident," Spencer said the next day. At 12:01, Masursky and some other team members were having a nightcap at the Loch Ness Monster, a nearby pub that was a favorite retreat of the Voyager scientists; they went straight home afterward without hearing the news, and didn't hear it until the next morning. Bradford Smith, though, was "blown out of bed," as he put it, by a phone call from the project manager's office at 2:15 A.M. Torrence Johnson found out two hours later, when Soderblom called him, and couldn't get back to sleep afterward. Andrew Ingersoll heard the news on his car radio on the way to work at eight, and his first concern, he later told me, was for the high-resolution pictures of the shaded underside of the rings. "The rings are, to me, the most unusual feature of the mission, and I was more worried about them than about my own stuff—the high-resolution pictures of Saturn's southern hemisphere," he said.

One of the last people to find out was Candice Hansen, who had walked in the door at 9:15 in the morning to be greeted with the news; she had spent the night sleeping in her car in the parking lot. "At twelve-fifteen last night, when I heard that Mission Control had reacquired the spacecraft's signal, I went to my car to drive home, but I was so tired that I crashed—that is, I sacked out right there, in the parking lot," she said. "I had my beeper—they could have beeped me, but they didn't." Their not doing so worked out well, for as she had been responsible for targeting the satellites—acting as the imaging team's liaison with the engineers—she would be most valuable when the pictures recorded during the occultation period came back.

Now, behind the closed door of the interactive room, she was closeted with Anne Bunker, who had the same role in targeting the rings as well as the atmospheric features on the planet, and a few other imaging-team members. They were watching the playback of the recorded data to try to see where, when, and in what manner the aim had gone wrong. In the meantime, the team quarters, as well stocked with rumors as an army barracks, was abuzz with hearsay: the platform had inadvertently aimed at the Sun, and all the instruments on it were ruined; the spacecraft had been hit by a meteor; the cold temperature in Saturn's shadow had frozen the platform's lubricants, and everything would be all right once the spacecraft had warmed up in the wan sunlight. Verner Suomi said he bet the high work load of the encounter period had caused the trouble.

Terrile and his colleagues in the adjacent interactive room—the group included Morrison, Smith, and Cuzzi—were watching the same recorded pictures as Hansen, Bunker, and the others next door. A series of eleven pictures of the rings was coming in now; it had been taken the previous night, between 9:17 and 9:33, not long after the Enceladus series I had seen, which had been the last to return directly from the spacecraft. The swirl of the rings was spectacular, with grooves and ridges in endless succession. "This picture will be the cover of *Science*," one observer said; and when it was succeeded by another just as good, someone else said, "This will be the cover of *Nature*." (As *Science* is in this country, *Nature* is the most widely read scientific journal in England.) Terrile pointed out, on a wide-angle picture, one of the F-ring shepherds and said that he thought it looked off center. "Everything seems to be okay through nine-twenty-five last night, anyway," he said, looking at a chronological list of the pictures. "I hope everything hung together long enough to get the ring-plane-crossing picture—the one of the rings edge-on." The next picture was a blank. It was supposed to have been a narrow-angle picture of the same F-ring shepherd, which, Terrile said, "would have been a great shot." The narrow-angle pictures were, of course, more affected than the wide-angle ones. Next came a wide-angle picture of the B ring lit obliquely by the rays of the Sun, which threw the ringlets and grooves into such relief that Cuzzi, in the back of the room, whistled in enthusiasm. Then came a picture of the outer border of the A ring. "Look at that detail! Gosh, that's pretty!" Terrile said. Another scientist said, "The pictures are so wonderful it's hard to realize they're going to end. Each picture coming in now might be the last."

Terrile, who knew more about the targeting than anyone else in the room, said to me, "Even though they're all great pictures, the pointing is off on all of them. It's desperately grave." The next picture, a narrow-angle one taken at 9:37, demonstrated, even to those unfamiliar with the targeting, what he meant. It was of the F ring, which in this shot was supposed to cut diagonally across the center of the frame, from the lower left-hand corner; instead, just a tiny segment of the ring was visible, across the upper left-hand corner. Terrile shook his head. "All the good

Later, when the Image Processing Laboratory had enhanced the F-ring segment even more, it was seen to be divided into five or six separate strings, glowing like fluorescent tubes.

stuff—the ring-plane-crossing picture, the encounter with Enceladus, the six-picture mosaic of Tethys—comes after this," he said. Despite his discouragement, however, Terrile was able to rescue something from the worsening situation. Enhancing the small section of the F ring, he said, "Hey, look at that! We've got lots of structure in there! I see three—no, *five* strands! Maybe the spacecraft got hit and it's wobbling, or maybe there's one strand with four residual images, like what you sometimes get on your TV set at home." He called Bradford Smith over; between them, they established that at least this segment of the F ring was definitely made of several strands. (Later, when the Image Processing Laboratory had enhanced the picture even more, the segment was seen to be divided into five or six separate strings, glowing like fluorescent tubes; all were rigidly parallel to each other, though some were probably on different planes, so that together they would make a three-dimensional structure.) In the interactive room, somebody remarked, "It's amazing how the discoveries keep right on coming, even after the spacecraft is incapacitated."

The F ring from 64,000 miles, as Voyager approached the ring plane. Picture enhanced several days later by the Image Processing Laboratory.

Some time later, in the conference room next to Bradford Smith's office, Candice Hansen told me what she and the others behind the closed door, with their greater knowledge of the targeting, had been able to figure out. She had tacked to the conference room's bulletin board hard copies of all the pictures taken since seven the previous evening. The pair of Tethys pictures that I had seen come in around eight-thirty were a little off center; the subsequent ones, though—two of Mimas that had come in five minutes later and the seven of Enceladus beginning at 8:42—had been right on target. A pattern began to be apparent—the aim going off, getting back on target, and then going off again in ever broadening swings. Soon it was noticeable that the aim was going off a little, and then a little more, but with no apparent pattern other than a general worsening of the situation. The high-resolution camera missed the two encounter pictures of Enceladus entirely, as earlier Terrile had feared they would.

At the bulletin board, Hansen told me, "We oscillated back on target for the ring-plane crossing, but—would you believe it?—the tape in the recorder that was storing the pictures came to its end and reversed as the data for the middle of this picture were being recorded, so that the one shot that would have given us an edge-on view of the rings had a

An image made August 26, 1981, just before the ring-plane crossing—and just before the dropout of data. The spacecraft was 64,000 miles from the rings, its wide-angle camera raking obliquely across the leading ansa. From the bottom up, there is the F ring, the A ring with its big gap, the Cassini division (the narrow dark band at the center), the B ring, and the C ring; they reverse toward the top, though the F ring is not visible at the far side. The white streaks in the B ring are spokes, white in the forward-scattered light.

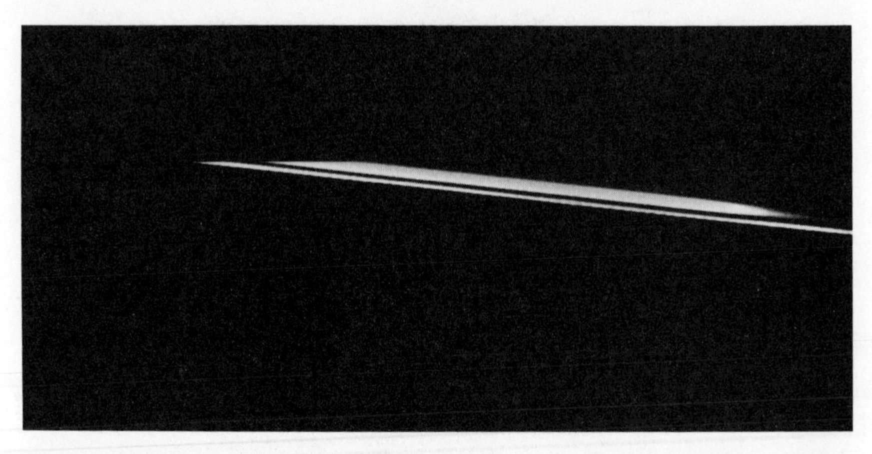

The ring-plane-crossing picture, with the data dropped out: The rings should continue to follow the line tilting slightly upward until they reach the left-hand margin—but because the tape in the on-board recorder reversed inopportunely, they are sliced off in a straight line parallel to the top margin. The picture was taken within four seconds of the moment of crossing—at no more than forty kilometers above or below the ring plane. Most likely the spacecraft was below, in which case the broad streak on top is the F and A rings combined; the black sliver is the B ring, dark beneath the ring plane; and the thin white streak below is the A and F rings, on the far side of the ansa.

data dropout right in the center! We lost the critical part of the image." A large piece of the rings—they were just a sliver now—was missing. (Many months later, something was salvaged, for it turned out that what is called a residual image—like what causes double images on television screens—of the critical picture had been retained in the camera's vidicon tube, and after much painstaking work was coaxed out of the data bits of the next picture. It was, however, too faint to be of much use.) What was to have been the six-picture high-resolution mosaic of Tethys came next; the camera had missed Tethys altogether on all but the last of the six pictures, which showed, in one corner, a tiny piece of one hemisphere. It was a jumble of craters, in such beautiful resolution—1.7 kilometers—that some of the scientists were all the more distressed at what they had lost. "I wanted to cry," said Hansen. "But I told myself that professionals don't sob in front of their colleagues. Then I got to working so hard trying to figure out what had happened that I stopped thinking of the implications." The cameras were next supposed to slue away from the moons and rings and toward the dark side of Saturn to look for lightning or auroras; as all the pictures were black, there was no way of gauging their aim—either they were on target but saw no light, or else they were off target. Then came some more ring pictures that were right on target—and, after that, nothing. "I suppose we should be grateful for the stuff we got," she said. "But this has been our work for years. It's hard to be objective about it."

Terrile, who was back in one of the interactive rooms, was in no mood to be grateful either. I asked him what he thought was the extent of the loss. "We've lost the highest-resolution mosaics of Enceladus and Tethys," he began, ticking off a catalogue on his fingers. "We've lost the stuff on the dark side of the planet, which might have revealed lightning or auroras. As far as the rings are concerned, we've lost the high-resolution scan of the underside, and without that we can't compare the top of the rings with the bottom, or with the underside of the rings as Voyager One saw them last November—we can't look for changes. And we've lost some F-ring images we could have used to compare its appearance over time or for stereo. It's a catastrophe."

As I headed for the morning press conference, I passed a group of scientists in the main hallway looking quietly at the television monitor suspended from the ceiling, where they could see the raws from the spacecraft still coming in—one black, empty frame after another. There seemed to be general agreement that it was a good thing that Admin-

The only image of Tethys that survived from what was to have been a six-picture mosaic at the extremely high resolution of 1.7 kilometers.

istrator Beggs had timed his visit and press conference when he had, and that he had returned to Washington before the midnight champagne celebration in Von Karman Auditorium.

Voyager's project manager, Esker Davis, had been in Von Karman Auditorium at the time of acquisition of signal at 12:01. He had some champagne, told the press that everything was fine, and left before the blank raws began coming down. Back in his office in the missions-operations building, he watched telemetry displays—the data that gives information about temperatures, pressures, voltages, and so forth—until a quarter to one, and, having seen nothing to disturb him, was just turning the lights off to go home when the phone rang and a flight engineer told him, "We've got a problem."

Richard Laeser, Davis's deputy as well as mission director, was called shortly afterward at home, just as he was going to bed. He got to his office a few minutes later, where he joined Davis and the other engineers and flight controllers, who were studying Voyager 2's telemetry. The telemetry data contained certain code numbers, called omens, that alerted the flight controllers to any unscheduled action taken by the onboard computer. That night, when the spacecraft came out from behind Saturn, one of the flight controllers had noticed, simultaneously, an omen and the fact that the platform was not pointing where it should have been in azimuth (the elevation was correct). When the platform had been unable to slue to one of its targets after sixty minutes of attempting to do so, a timer had automatically turned off the motor that moved the platform in azimuth and had at the same time disabled the computer program for that target. There existed in the computer another program, which, in the event of such a failure, would command the platform to turn to a safe position, one where the instruments would be least likely to face the Sun, which would burn out their sensors. It was this unscheduled movement that had triggered the omen. Whatever had prevented the platform from reaching its original target, though, was also preventing the move to the safe position; accordingly, the platform's azimuth motor, over the next several hours, kept trying to slue the platform to the safe position, failing, shutting off, and trying again. The platform, which was at 90° of elevation in its vertical axis, was stuck in azimuth at 260°. Since at zero degrees azimuth the platform is pointing directly at the spacecraft and at 180° directly away, the platform was stuck almost three-quarters of the way through its complete rotation. "It was as if there were a stone wall there," Laeser told me later.

At 2:10 A.M., a command was sent to the spacecraft—the first since the accident—to accomplish in elevation what could not be done in azimuth; that is, to turn the instruments to a position safe from the Sun. Five minutes later, Bradford Smith was called to take part in discussions about whether or not the cameras should be turned off; it was decided to leave them on but to reduce the electricity going to them and to the other platform instruments, in order to cut down on the damage in case they still somehow managed to face the Sun. While these instructions were being sent up, the photopolarimeter, which has the widest field of view, did see sunlight, triggering a safety mechanism that immediately shut it off before it could be hurt. (The safety mechanism was a last resort; no one had wanted it to be put to the test.) Finally, at 5:10 A.M., a command was sent up disabling the safety program, which was still pushing against the stone wall, trying to move the platform in azimuth; should it suddenly succeed in moving, it might inadvertently aim the instruments into the Sun.

At about the same time, another command was sent up: to back the platform, in the lowest of its four gears, away from the stone wall. When the results of this maneuver came back later that morning (one of the biggest problems was the almost three-hour round-trip time for communications), the flight controllers found that the exact opposite had occurred, because of what Laeser called "an idiosyncrasy" of the spacecraft. "We don't like to talk about it very much," he told me later, going right ahead: Flight controllers give instructions to the platform on the basis of their readings of one of two instruments called potentiometers, both of which tell where the platform is in azimuth and how much current will be needed to drive it to another specified position. Because of what Laeser called "an eccentricity of design," the two instruments give opposite readings; consequently, the flight controllers have to know which of the two they should rely on—an assessment that normally requires several hours of analysis. With the spacecraft receding from Saturn at 23,474 miles per hour, the engineer doing the analysis was working quickly, and he made a wrong assessment. The result was that the platform, instead of backing up, moved forward in the lowest gear six degrees, at last running right over the stone wall at 260° azimuth and stopping at 266°. "It was a fortunate mistake," Laeser said. "That was the first hint we had that we weren't dealing with a stone wall after all." They could now begin to plan gentle maneuvers in the lowest gear in order to test the platform's mobility.

Even as the spacecraft hurtled away from Saturn, no one raised the possibility of attempting to salvage the pictures and the other scientific information by trying to ram the platform back to face Saturn or by rotating the entire spacecraft in order to aim the instruments. The former course would have risked ruining the platform for good, and the latter would have required the writing of new computer programs, which would have taken many days; indeed, the transmission time to and from Saturn in itself was enough to rule out any kind of speedy solution to the problem. Moreover, budget cuts had reduced Voyager's engineering staff from what it had been two years previously, at the time of the two Jupiter flybys, so there were not enough engineers to provide the kind of fast analysis in depth needed to correct a failure during an encounter period; the engineers would first "safe" the spacecraft, as they put it, and then work on the problem as fast as they could with the resources they had. Several months earlier, Davis and his staff had decided that the objectives at Saturn had been accomplished to such a great extent by Voyager 1 that Voyager 2's highest priority would be not Saturn but Uranus. If the platform had broken down a month before the encounter, instead of halfway through it, the procedure would have been the same. Not the least amazing aspect of the Voyager mission was the matter-of-fact way in which the project's managers decided to cut their losses at a planet nine hundred million miles from the Earth, in order to explore the next, two and a half billion miles farther away, and a third nearly two billion miles beyond that—a choice that was presenting itself for the first time in the history of mankind. Edward Stone's principle of basing decisions on what would provide the most new science was being pushed to the farthest reaches of the solar system.

Wednesday morning's press conference was, not surprisingly, unusually well attended. Richard Laeser, who had been up since eight o'clock the day before, and Davis, who had gone home at six for a few hours' sleep and had returned looking singularly unrefreshed, spoke of the previous evening's events, which were not yet fully understood; the telemetry data recorded while Voyager was behind the planet was still coming down and wouldn't be completely in hand for several days. Even so, the reporters bombarded Laeser with questions about what might have happened, such as whether the spacecraft had been hit by a chunk of ice or bombarded with particles during the ring-plane crossing. He couldn't exclude this possibility, Laeser said; the possibility was rendered more likely in the minds of some of the reporters by a recording of radio noises picked up by a pair of long antennas aboard the spacecraft and played at the press conference. The recording was a series of hisses, grindings, and thumps; some of the noises, Frederick Scarf, the leader of the plasma-waves team, said, might have been made by ionized dust particles hitting the antennas.

Laeser was followed by Stone, the chief scientist, who said that even without the platform instruments Voyager was still very much in business, since the spacecraft's other instruments were sending back data as usual. He then ran down the list of pictures that had been missed, in a manner distinctly more upbeat than Terrile's: While the highest-resolution mosaics of Enceladus and Tethys had been lost, the satellite observations had otherwise been completed. While the close-up high-resolution imagery of the underside of the rings had been lost, that of the bright upper side had been accomplished. They had, of course, lost some high-resolution F-ring pictures, but these were mostly of places that had already been covered from another angle. While there was a strong likelihood that they would miss some of the high-resolution imagery of the southern hemisphere of Saturn, a certain amount of this had been done by Voyager 1. "It's fortunate that the platform didn't stop a few hours earlier," he concluded. "We have completed most of our observations. The mission is already a success. We have achieved all mission data."

The meeting of the imaging team at one o'clock that afternoon was pervaded with gloom. A television monitor in the far corner of the conference room dutifully continued to display one blank, black raw after another. In an attempt at lightheartedness, a couple of younger team members had attached to the end of the series of pictures of Hyperion a picture of Phobos, a Martian moon that was similarly potato-shaped, and now passed the series around the table. Carl Sagan and Harold Masursky, both of whom remembered Phobos quite well from the Viking mission to Mars, immediately spotted the hoax, but neither of them cracked a smile. Someone handed around a copy of the ring-plane-crossing picture with the data dropped out of the middle. "Just like the mission," someone else commented. The geodesist Merton Davies said, "We're missing our high-resolution mosaics of Enceladus and Tethys, in addition to the other things. We shouldn't minimize the loss to the press," to which Laurence Soderblom, who was chairing the meeting, replied, "Even with the losses, we have still managed to characterize all the satellites in the Saturnian system."

Clearly, there were two ways of viewing the situation, and which one a particular scientist adopted—whether he concentrated on how much milk had been spilled or on how much still remained in the glass—depended in part on how his own pet interest had been affected. Moreover, since it was obviously better for NASA, and for the scientists themselves, not to have a disaster on their hands at a time when the funding for future missions was at risk, there was a certain insistence in the arguments of those scientists who wanted to play down the situation, and this annoyed the scientists who were suffering real losses. The loss of the two Enceladus pictures with a resolution of 1.6 kilometers (as opposed to the seven they had received with a resolution of 1.9 kilometers) was perhaps not as great as the loss of the six Tethys pictures at 1.7 kilometers, because the best pictures in hand of that moon (the tiny segment aside) had a resolution of no better than five kilometers. The loss of coverage was serious in the case of both moons, however, for now neither of them could be measured all the way around with the precision that Davies would have liked, or mapped as comprehensively as the mission cartographers had expected. (Harold Masursky said later that day, "We have, as Soderblom said, gotten a sample of all the satellites. But the fact that we got only one of three mosaics of Enceladus and one of two of Tethys, hurts. These are painful things to miss. We've fought for every one of these observations for months, not only within the imaging team but with other teams as well. Some of us don't like to see the whole thing dismissed as a 'successful mission.' ")

Eugene Shoemaker, an optimistic man who likes to concentrate on what's in the glass instead of what's on the floor, said that he had enough Enceladus pictures to begin making a thermal gradient of that moon. "We were hoping to see a few craters on Enceladus with topographical relaxation or distortion because of some warmth in the satellite and a resulting softness in the crust, and we see a whole panoply of them," he said. "The larger craters tend to distort and dissipate first, of course." He held up a picture of Enceladus and, pointing to a flat, relatively smooth area, went on, "You can see that the smaller craters—ten kilometers or so in diameter—are crisp, while the craters in the thirty-kilometer size range are very much relaxed. There *are* no craters bigger than that, which indicates that those have relaxed away altogether. Now we'll apply a heat gradient I worked up for Ganymede, and for the time being you'll just have to take it on faith that it will work on Enceladus—the surfaces of the two moons are very similar." Assuming the same heat

The Enceladus mosaic processed, in black and white, for contrast and relief, to bring out the variety of terrain and the distortion of the craters—both indications that the satellite had been warm.

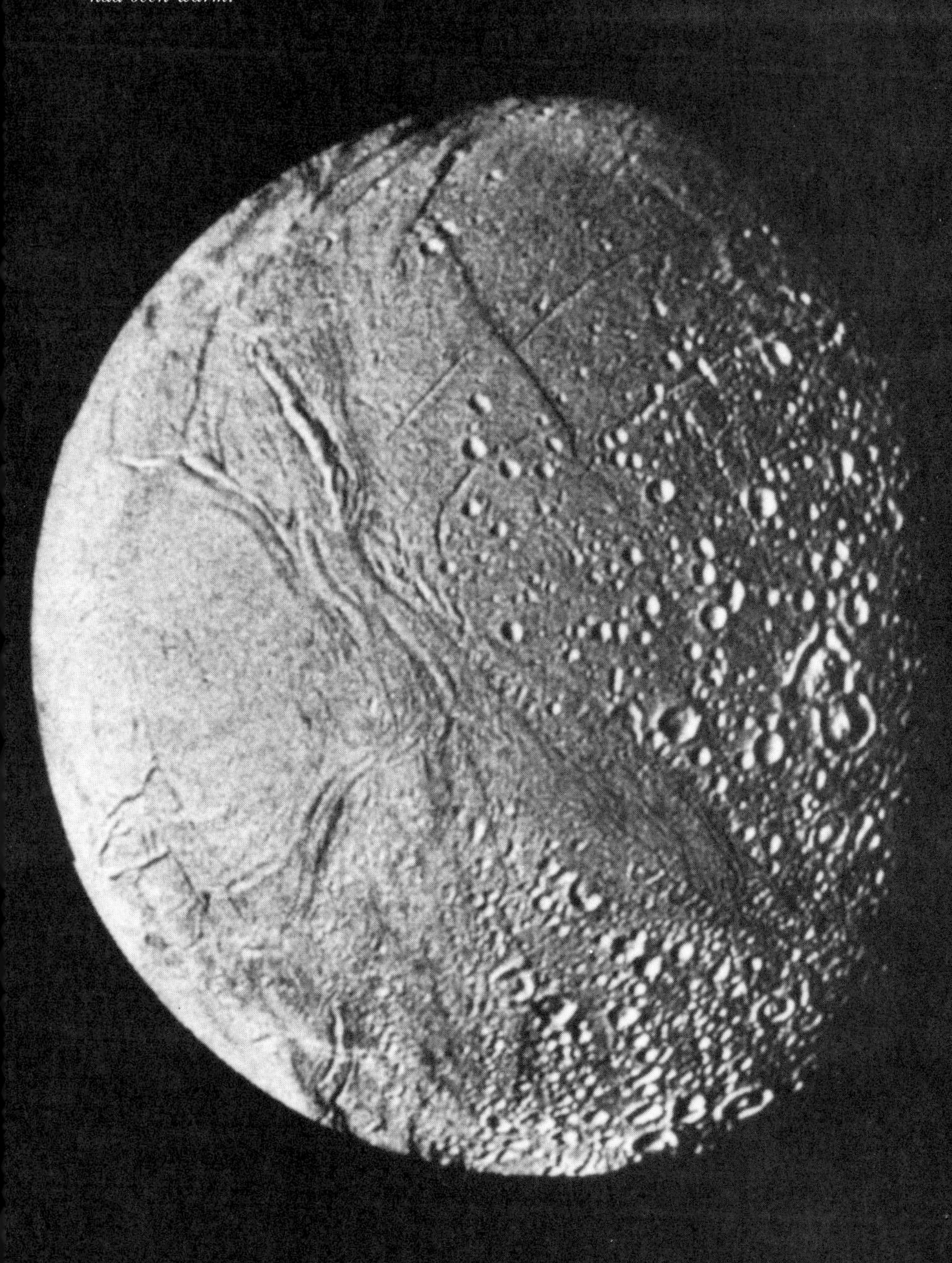

gradient as Ganymede's, Shoemaker went on, one would expect Enceladus's temperature to rise 5° K per kilometer, moving inward, and therefore at the particular place where he was looking, Enceladus would be liquid several tens of kilometers down.

(Shoemaker and a graduate student, Quinn Passy, revised these figures somewhat later, when it turned out that Enceladus's cratering was not as much like Ganymede's as Shoemaker had first thought. However, as the composition of the crust was also different from what had been thought—the ices of Enceladus are now thought to contain some ammonia, which has a far lower melting point than water—the results are, by fluke, very close to Shoemaker's estimates at the time of the encounter: Enceladus would be liquid a little nearer the surface, between ten and twenty-five kilometers down; but the temperature would increase at the same 5° K per kilometer. The ammonia could well have been mixed in with the water ice composing Enceladus as clathrates, in a manner similar to what may have happened on Titan and possibly other satellites of Saturn as well. In the opinion of David Stevenson of the California Institute of Technology, whose theory it is, the lower melting point of ammonia ice would not only mean there would be liquid nearer the surface, but would also mean that crater distortion and flooding would occur more readily, at temperatures a hundred degrees lower than previously thought. Consequently, Stevenson and a colleague at Cal Tech, Jonathan Lunine, also believe that Titan may once have had an ammonia ocean a hundred miles deep, and may still have volcanoes vomiting a water-and-ammonia magma. Whatever the case with the ocean, the flooding, or the liquid layer beneath the crust, Shoemaker describes the fluid as "an ordinary detergent, water and ammonia, good for cleaning your floors.")

Enceladus was complicated, Shoemaker said at the team meeting, for there were regions where even craters with diameters in the ten-kilometer range had relaxed away, and in those spots, Enceladus was, at the time the relaxation took place, presumably liquid even nearer the surface. (This would be true with or without the ammonia.) Very likely, such a region had retained its heat since the moon's initial formation. In still other areas, he had found evidence that the heat had lessened with time; this suggested that the eccentricity of Enceladus's orbit—causing the alternation in the tidal force of Saturn's gravity as the moon swung closer to and then farther from the planet—had become less pronounced; indeed, such an eccentricity would tend to dampen in time.

Clearly, if the tidal-energy theory was to explain the appearance of Enceladus's surface it would have to embrace variations in heating in time and place. Later, Charles Yoder, the JPL scientist who had first proposed that Enceladus was hot, amplified his theory, suggesting that the satellite was, early in its history, a rigid, unfractured body, and so the tidal energy from Saturn, unable to cause frictional heat, because there were no faults whose edges would rub together, went instead into pumping Enceladus's orbit into greater and greater eccentricity, which in turn increased the tidal stresses on the crust, until at last the satellite cracked. At once, the energy began to dissipate in frictional heat, all the greater because of the suddenness of the event; the ice along the cracks got slushy, perhaps managing to flow out onto the surface before it either evaporated or refroze. Certainly, parts of the surface look as if they had been flooded. As more tidal energy went into heating Enceladus, less went into pumping its orbit out of shape; as the orbit gradually became nearly circular, the tidal energy was reduced, the satellite cooled, cracks refroze, and the body became rigid once more: then, of course, the cycle would start all over again. Enceladus, according to Yoder, would have undergone several such pulses (as he called them) of heat over its four-and-a-half-billion-year history. His theory seemed to explain not only Shoemaker's observation that Enceladus had been hot at different times in its history but also his observation that the effects were not uniform across the moon's surface, for during each period of tidal stress Enceladus would probably have cracked in different places, and it would be along these cracks, of course, that the heat would be greatest.

After this meeting, most of the imaging team moved upstairs to the joint science-team meeting, where Shoemaker repeated his Enceladus report. Scientists for the other platform instruments reported that they had all lost observations. The ultraviolet and infrared spectrometers had both missed scans of the planet; the infrared spectrometer had missed shots of some moons going in or out of Saturn's shadow; and the photopolarimeter had missed the occultation of Beta Taurii by the rings. There was, however, still plenty of material. The photopolarimeter team had managed—before the breakdown of the platform—to measure the brightness of Delta Scorpii as its light passed through a number of the rings. The photopolarimeter on Voyager 1 had broken down during the Jupiter encounter and had been the only instrument on that flight to return no Saturn data, and this fact heightened the team's success now, to say nothing of its earlier excitement when Delta Scorpii had begun to blink on and off at various intensities as it disappeared behind the rings and passed the various ringlets and gaps—or grooves and ridges, if that is what they were. The instrument had scanned some forty-four thousand miles across the F, A, B, and C rings, at five-hundred-foot intervals: the team's computer printouts, if laid end to end, would cover half a mile. Moreover, the photopolarimeter's resolution was about seventy times as good as that of the narrow-angle camera, whose best resolution of the rings had been ten kilometers. The photopolarimeter team had been able to look at very little of the data before the joint science-team meeting, but already they were able to throw light on the important question of how thick the rings are. The best guesses so far had been made from the Earth, on those occasions when our planet crossed Saturn's ring plane; when this happens, the rings appear as a thin line, which some astronomers estimated to be only a mile deep. (It had been hoped that the ring-plane-crossing picture taken the evening before would refine this estimate, but doing so was no longer possible now that the crucial part of the picture was missing.) The photopolarimeter had not looked at the rings edge-on, but the team had been able to make a guess at the thickness of the A ring, since the data showed how long it took Delta Scorpii's light to go from full intensity, before its occultation by that ring, to its lowest intensity, when it was veiled by the ring; the theory was that the longer that took, the more the edge of the A ring must taper, and therefore the thicker it must be. The transition from maximum to minimum light intensity was about as abrupt as the instrument could measure, spanning three data points, or about a thousand feet, and the model that these data seemed to fit best showed the A ring to be about as thick as the distance between two of the data points—five hundred feet, or only about a tenth the thickness estimated from the Earth.

The first commands to test the scan platform in the lowest gear had gone up to Voyager 2 at one o'clock that afternoon, and the results would be analyzed well into the evening. The command had been to move the platform back from 266° (where it had come to rest after knocking over the stone wall at 260°) to 255° and then forward again to 270°; when the results came down, they showed that the platform hadn't made it all the way back to 255°, but it had made it all the way forward to 270°. This was encouraging, for it seemed to confirm an impression that Davis, Laeser, and many of the engineers had had that morning—that the situation was reminiscent of a platform breakdown on Voyager

1 not long after it left the Earth four years ago. That breakdown had been caused by a particle of insulation inside the azimuth gearbox that had been caught between a couple of gear teeth, jamming the motor and hence the platform. After months of evaluation, using identical equipment on Earth, engineers were able to duplicate the telemetry from the spacecraft so exactly that they could identify the particle and its size. They also found that they could overcome the obstacle by moving the platform back and forth in the lowest gear until the bit of insulation was ground away. Still, Richard Laeser was not entirely happy with this diagnosis for Voyager 2. "If there had been any loose material in Voyager Two's gearbox, it would have got caught long ago, and not waited until four years after launch," he told me in his office a few days later. "Also, there is the coincidence of the timing, which seems to suggest that the problem had something to do with the ring-plane crossing." However, the engineers didn't have many other theories, so late that night they decided to repeat the earlier test, grinding back and forth over the obstacle once again; if that went more easily, they would think in terms of making longer and longer swings.

When I arrived at the imaging team's quarters Thursday morning, things seemed to be looking up. The engineers had learned that the commands sent up the previous night to pass back and forth once more over the obstacle had been carried out successfully, although the backward leg—from 270° to 255°—had been accomplished at only a third the expected speed. Before Davis and Laeser decided on the next step, though, they sent everyone home to bed; a number of engineers had been up for forty hours, and their thinking was not as sharp as it had been the day before. They would be back early that evening.

At the morning's press conference, Edward Stone said that he had given some further thought to a question asked the previous day about what percentage of success he would put on the mission, and he now felt that the mission had been *two* hundred percent successful.

Most of the scientists, however, were in a worse mood than they had been the day before. At the imaging-team meeting that afternoon, Laurence Soderblom, who was once again the chairman, tacked up a photocopy of the seven-picture high-resolution mosaic of Enceladus that Voyager 2 had sent back before the mishap; he pointed out where the two pictures of Enceladus that the cameras had missed would have been. "What we missed was around here, in the ancient, heavily cratered area, so they wouldn't have told us anything about the recent evolution of Enceladus," he said.

"It's lucky we didn't get them, then," said Harold Masursky, who was clearly opposed to making things seem better than they were, even with the most sugarcoated of sour grapes.

Next, Soderblom had an announcement with regard to the Public Affairs Office's televising the reactions of members of the imaging team to the incoming pictures. "PAO is redoing the televising schedule, as the data coming in aren't worth analyzing real-time," he said.

"Or any other time," Tobias Owen remarked, staring at a blank raw on the monitor.

"Is there a report from the Iapetus working group?" Soderblom asked, trying to move on.

James Pollack reported on that group's attempts to establish whether the black material on Iapetus had come from inside or outside the satellite. "It is safe to say the stuff is either exogenic or endogenic in origin—or both," he said. On this negative note, the meeting adjourned. Even by the following January, when the imaging team's report appeared in *Science*, the question of the provenance of black material on Iapetus had not been resolved.

The pervasive gloom had not seemed to dampen Richard Terrile's spirits. An excited yelp from his office brought me hurrying in. "Want to see something hot?" he asked. "Cuzzi and I asked the Image Processing Laboratory to blow up to the same size pictures of opposite sides of the rings, so that we could study individual rings as they went all the way around the planet. Look at these two pictures, taken diametrically across the rings from each other, about four hundred and seventy-five thousand miles apart, where the Cassini division meets the B ring and right next to the gap that's supposed to be the Mimas two-to-one resonance. Now take a look at the outer edge of the B ring in both places. Isn't that wild?" He had cut the two pictures from one side with the left half of the picture from the other side, the rings and gaps of the Cassini division lined up exactly, in a demonstration of perfect circularity, but the outer edge of the B ring did not line up nor did the next several ringlets within it. Something was warping the outer edge of the B ring; moreover, whatever was doing it was different from the counterbalancing forces of gravity and resonances that caused the density waves. Terrile, who generally favored resonances over embedded moonlets (of whose exis-

Something was warping the outer edge of the B ring: When two pictures from opposite sides of the rings are matched up, the A ring (right) fits, and so does the outer edge of the Cassini division; its inner edge, and the outer edge of the B ring, do not (though, moving inward, to the left, the B ring's structure realigns once more). Richard Terrile wondered if the effect might be caused by the Mimas resonance.

tence he was increasingly doubtful) as the major influence on the rings, was excited by the possibility that the warping might have to do with the Mimas resonance alone. He would have to find out precisely where on its orbit Mimas was when the pictures were taken, he said, for if the warping had been caused by Mimas it would travel around the ring synchronously with Mimas. "Here we've been worrying about those tiny embedded moonlets, while right under our noses all the time there's this much bigger effect," he said.

Terrile had an appointment in a few minutes at the Image Processing Laboratory, to view a second film made of the spokes on the B ring several days ago, and he invited me to come along. Unlike the earlier spoke film he had seen, this one was taken when Voyager was much closer to Saturn, and instead of following groups of spokes all the way around, it focused only on the leading ansa—the sweep of rings coming around the left-hand side of the planet—thus catching far more spokes moving through a far smaller area. The two films together were giving a new picture of the evolution of a spoke.

As we walked up the central mall to the laboratory, Terrile told me how he thought the spokes changed as they moved around the B ring, as best he could figure it out from looking at the previous spoke movie and some stills of this one. "They start off looking like thin, scratchy pencil lines appearing radially across the B ring, straight out from Saturn, as though whatever caused them acted like a beam coming from the center of the planet," he said. "Then, as they get older, they get pulled into a smudgy wedge shape by what we call Keplerian shear—simply the result of the fact that particles nearer the planet are orbiting Saturn faster than the ones farther out. We were not sure this was happening at the time of Voyager One's encounter, but we now know that it is. The wedges, whose thick ends are toward the planet, get smaller as they move out, until they come to a point; then, if they keep extending outward, they widen again; consequently, a full-grown, adult spoke looks something like an hourglass. If, as I suspect, the pinched part of the hourglass is at the spot where the orbital speed matches the speed of the rotation of the planet—what we call the co-rotational point on the rings—it would confirm a theory we had earlier: that Saturn's magnetic field, which rotates at the same rate as the planet, has something to do with causing the spokes. I haven't had time to make the measurement yet."

Terrile hurried into the Image Processing Laboratory, a big concrete building across the court, and slammed the door behind him. "It's cool in here," he said. "Where there are computers, it's always cool—from the air conditioning for the electronics." After standing for an instant in the lee of a large whirring console, he ran up three flights of stairs and into a small, square gray room containing several television sets. On one was the laboratory's enhancement of a picture of the spokes taken just before Voyager's ring-plane crossing, with the Sun almost directly beyond the rings and shining across them toward the cameras, an angle at which fine particles, such as the spokes were thought to be composed of, would best be brought out. In the picture, a couple of spokes could be seen—though whether or not they were suspended above the B ring was impossible to say.

The technician took us into another room and ran the new spoke movie for us. Terrile said that this new film would be more useful than the previous one for close measurements of the spokes; he could already see that they didn't extend outward as far as he had thought, and that they began a good deal farther in toward Saturn. (Several weeks later, after Terrile had studied the frames in detail and made measurements, these impressions were confirmed, as well as his earlier impression that the pinched part of the hourglass coincided with the co-rotational point on the rings. Moreover, when he plotted several of the blurry wedges backward to their starting points he found that the spokes were born randomly all around the B ring—though when I talked with him still later, he thought there might be a nonrandom component, too. Clearly, the spokes would remain a puzzle long after Voyager 2 was well on its way to Uranus. As the spokes rushed around what looked like a section of the rim of a plate, they resembled more than anything else a ghostly parade of circus animals. Some of the wedges had what looked like feet and long, tapering necks like giraffes (a few, in clusters, seemed like whole herds of those animals); others had sharp horns, like ibex; there were a couple of unicorns. From this close-in perspective, it seemed an oversimplification to call them wedges at all.

Four frames from the latest spoke movie, progressing from top left to lower right: a parade of ghostly giraffes.

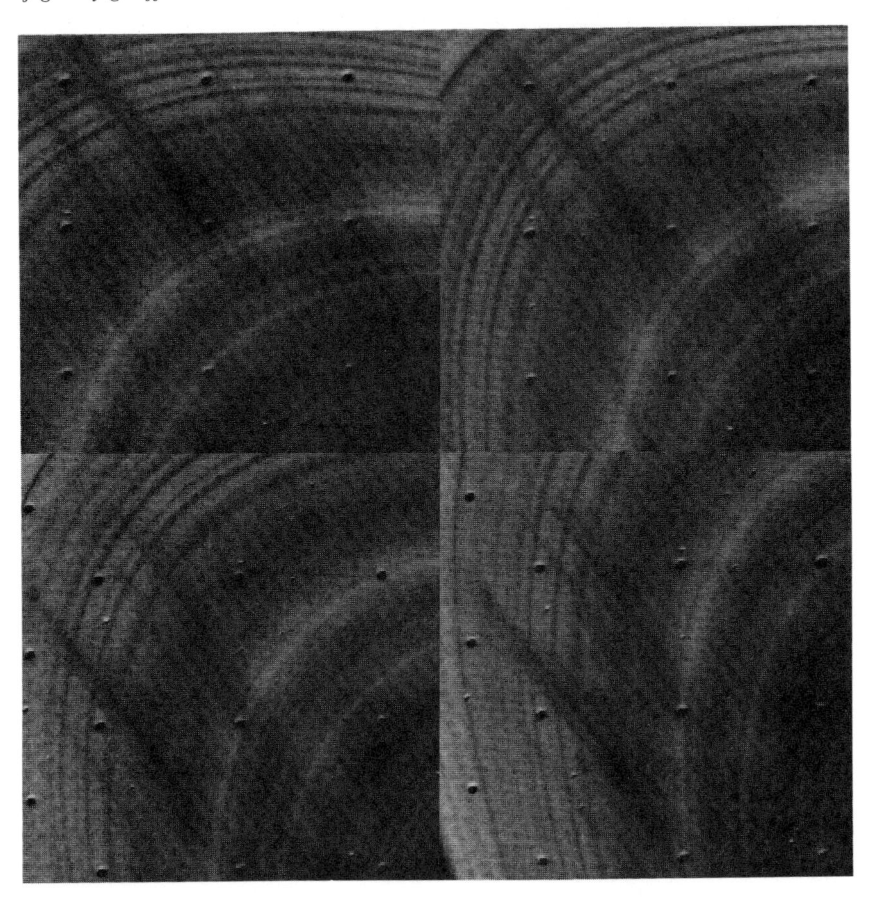

When the engineers and flight controllers got back to work Thursday evening, they were so encouraged by the result of the latest back-and-forth swing over the remains of the stone wall that, rather than repeat the same movement, they decided to command the scan platform to back very slowly from 270° across the obstacle, and on to 209°, which would aim the instruments back at Saturn. This was also within the range that would be used for looking at Uranus, so if the platform were to break down completely, 209° would be a good place for it to be. Backing all the way there in the lowest gear would take many hours: with luck, the first pictures of Saturn should appear on the television monitors at JPL later the following afternoon.

On my arrival at the imaging team's quarters the next morning—Friday, August 28th—things were decidedly brisker than they had been for some days, for word had reached the scientists that the platform was slowly inching back to face Saturn. "We may have pictures in six hours," Torrence Johnson said, as cheerful as he had been on the morning before the encounter.

In the corridor outside the interactive rooms, Edward Stone was in animated—even heated—discussion with Andrew Ingersoll and Rudolf Hanel, the head of the infrared-radiation team; they were arguing about where, exactly, the platform instruments would be aiming in elevation when the platform had been turned back to Saturn. Much of Saturn, including the center of its disc, would be in shadow. This did not bother the infrared spectrometer, which was sensitive to heat rather than light; indeed, it was in the dark center of the disc that Hanel was most interested in making measurements. Ingersoll, however, wanted to aim nearer the limb, where there would be sunlight; a picture of the center was useless to him. "We had to go through negotiations in five minutes that six months before, when we were doing the targeting, would have taken weeks," Ingersoll told me when Stone and Hanel had left. "Stone was saying that the engineers have been working hard, getting very little sleep, and the scientists should not be too tough in their demands. I agree. The engineers are the heroes. It's up to Stone to talk to us and then pass on one single recommendation to them—poor fellow! I'll be happy with anything I can get. It happens that at the distance we are now from Saturn, a wide-angle picture will cover the whole disc. It's the narrows—the high-resolution images—that I'm interested in, though, and in the center of the planet, where Hanel wants to look, they're no good, because it's dark there. I have to target toward the south, where there is a lighted crescent. I think we'll use the wide-angle camera, which won't be as much use as the narrow-angle. The scan platform will be parked at two hundred and nine degrees—but there's a chance they can target the narrow-angle camera to the interesting parts by reorienting the spacecraft. That would be fantastic! We are so eager to get the cameras on target again! This is our one chance to see clearly the wind patterns in the southern hemisphere."

Though almost three days had passed since the failure of the platform, the daily press conferences were as lengthy as ever, partly because most of Voyager's instruments were still returning information, and partly because all the instruments, whether or not they were on the platform, had already returned such a wealth of data that there was a considerable backlog. The breakdown of the platform, followed by, if anything, an increase in the stream of information coming from the press conferences, demonstrated that in science it is in the computer or laboratory work, and not so much in the exploration or fieldwork, where events really unfold. Indeed, the IPL, with its computers that got the most out of pictorial evidence, was like a whole second spacecraft that took its own trajectory through the data, making new discoveries at its own pace. For instance, Arthur Lane of the photopolarimeter team, which was still going through the half-mile of printouts of the rings, had just sampled his data on the F ring, which turned out to have far more structure in it than what the imaging team saw; indeed, when a computer in the IPL turned the peaks and valleys of the polarimeter data, seventy times as detailed as the imaging data, into pictures, which it could do, the F ring's major components broke into a great many more rings and gaps; instead of the four or five main strands the imaging team had seen, the photopolarimeter saw ten; and what the imaging team had seen as the single brightest strand—the fluorescent tube—turned out to have an enormous number of ringlets inside it, like a section of the C ring, say; there were even sets of receding density waves, invisible to the cameras. (Ultimately, in all the rings, it found close to thirty sets of density waves.) The computer had turned out the photopolarimeter picture in glowing oranges and crimsons, as gaudy as a jukebox, so that it never could be confused with a picture from the cameras.

One picture Bradford Smith showed at the Friday press conference had been taken eleven days before, when the spacecraft was still half a million miles from the planet; it had only just come from the Image Processing Laboratory. It was an enhanced color view of the rings, assembled from frames taken with clear, orange, and ultraviolet filters. The false colors that resulted indicated the presence of different chemical compounds mixed with the water ice that made up the ring particles. (In the opinion of most scientists, the previously observed differences in the rings were still due to the number and size of their particles.) Only minuscule amounts of chemicals were needed to account for the color differences. Apparently, the outer part of the B ring (turquoise) differed in its composition from the inner part (ocher); the Cassini division, though, was the same Wedgwood blue as the C ring; and the A ring, which was gray, differed from all the rest. Clearly, if a picture ten days old could still make news, it would be very easy to carry out an idea Torrence Johnson and Rich Terrile had had several days before, at one of the daily meetings to decide what pictures to release to the press: to hold the Saturn pictures hostage for the funding of future missions, releasing them, perhaps, one at a time, to whet the national appetite.

The F ring, in a computer image generated from the photopolarimeter data, which was a cross section made by the light from a single star. With seventy times the resolution of the best camera image of the rings, the single brightest strand observed by the imaging team turns out itself to divide into countless ringlets, complete with density waves.

This false-color picture, along with other evidence, had set Eugene Shoemaker to wondering whether the major rings, since their chemical composition differed, might be the remnants of three moons—"parent bodies," he called them—that had been smashed apart in the early days of the solar system by the bombardment of meteoric and cometary material. He gave me the benefit of his latest thinking on the telephone after I had returned to New York. Bombardment, as he had already told me, would be far greater near Saturn than farther out, since Saturn's gravity would focus such a bombardment on itself. The parent body of the B ring, Shoemaker said, was probably a little bigger than Mimas, and the parent body of the A ring a little smaller; the parent of the C ring would have been the smallest of all—about forty miles in diameter. All three of the parent bodies themselves were very likely remnants of a larger parent (or grandparent) body, which Shoemaker thinks was big enough to have enough heat to differentiate somewhat—that is, to have internal zones of somewhat differing chemistry, accounting for the slight color differences observed in the rings. Indeed, Shoemaker likes to account for the entire Saturnian moon-and-ring system by three orbiting superbodies, each about the size of Titan. He told me that he could account for all the material inside the orbit of Titan—the inner moons as well as the rings—by one such body, perhaps a little smaller than Titan, and all the material outside Titan by another somewhat smaller body; Titan itself—unfragmented—would be the third. He believes that Titan's size—roughly speaking, three thousand miles in diameter, similar to that of Ganymede, our Moon, and the planet Mercury—was the most common size of objects when they first aggregated from the solar nebula. Jupiter has four such moons, and Saturn and the Earth each have one, making a total of six still extant.

These ideas, of course, have put Shoemaker himself on something of a collision course with others, such as James Pollack, who believe that the rings are the residue of the primordial planetary disc, and that they formed during the final collapse that formed the planet itself, which in an early phase might have filled a volume out as far as the orbit of Tethys. Shoemaker insists that subsequent bombardment by planetesimals, cometlike bodies, and meteors would have obliterated the rings if they had been a residue. Pollack says that Shoemaker probably doesn't know what the early rate of bombardment was. And Laurence Soderblom, whom I also asked about the matter, says that although bombardment in the ring region was probably great, it may have abated before the rings were formed; after all, if the rings *were* left behind by a collapsing Saturn, they would have been the latest part of the Saturn system to appear. There, for the time being, the matter rests—in a not untypical, much-impacted body of contradictory ideas.

At the imaging-team meeting Friday afternoon, there was further discussion—and some dispute—on the subject of Enceladus. A number of scientists, including Tobias Owen, were questioning the idea that Enceladus had ever been heated tidally at all. Something that Charles Yoder, who had been refining his calculations, recently said was causing the trouble. "Yoder now says that Mimas should have fifty times the tidal energy that Enceladus does, yet Mimas is one of the coldest, most rigid bodies in the Saturn system; it doesn't show the variety of surface features that we have seen on Enceladus, nor does it have the signs of flow or distorted craters," Owen told me. "If tidal heating is doing the job on Enceladus, you'd think that Mimas would have the same signs and more."

When I saw Shoemaker, a partisan of tidal heating, after the team meeting, he said, "It's true that Mimas has fifty times the tidal energy that Enceladus has. A planet has to be fractured to be kneaded and heated tidally at all, of course, but it can be cracked too much. Mimas, being the innermost big moon, has been hit harder, and more often, than Enceladus, because of Saturn's focusing of the early meteoric bombardment. As I said earlier, I think Mimas has been blasted apart several times, with the pieces reassembling by their own gravity. The whole body may be just a pile of rubble, and in that case the heat from tidal stresses would be dissipated quickly. And there is no way to concentrate it. Enceladus is less rubbly; even though it has probably been broken apart, too, this wouldn't have happened as often, because it is bigger than Mimas and farther away from Saturn. The fact that it is less rubbly means that it loses heat more slowly. If Yoder's heat-pulse theory is correct, you might have had high heat on Mimas but you would have had episodes of even greater heat on Enceladus, and heat that was concentrated locally."

About two weeks later, Yoder calculated that the tidal energy going into Enceladus might not have been enough to keep cracking the satellite, and he substituted another process, which he called "creep viscosity"; here, the tidal energy would heat Enceladus in the manner that an iron bar is heated simply by bending it back and forth but without cracking it. As the heat could be built up and released rather quickly, a considerable amount of it could still be concentrated at one time and one place. (With the subsequent advent of David Stevenson's ammonia theory, less heat would be needed to account for the same signs of melting or distortion.) There would be the same pumping of the orbit to greater eccentricity and the same return to greater circularity, so that the periodicity of the old theory is retained. For his part, Shoemaker is sticking to the old theory in its entirety; if it's cracks Yoder needs, he says, he can crack Enceladus for him any time with giant impacts. Shoemaker feels that creep viscosity is very likely at work, too.

While I was walking back to the imaging-team quarters that afternoon after the science-team meeting, which followed the imaging-team meeting, I asked Masursky what he thought about the Enceladus controversy. Masursky, it turned out, had his own line of thought: "The outer planets, and their moons, have really changed our thinking," he said. "It used to be axiomatic that the bigger the body, the more internal activity it had, for it contained more radioactive elements, and its greater volume kept the heat in. Ganymede and our own moon show signs of having been hot—but then we saw Io, which is smaller than either of them, and it turns out to be far and away the hottest planetary body in the solar system. Io really reversed our thinking! It showed that other forces aside from radioactive heating could be involved in altering a body; in that case, the force was clearly tidal energy. And now here is Enceladus, many times smaller than Io, and it also is showing signs of heat. Now if tidal energy, or gravity, proves not to be responsible, could other fields—electromagnetic, or radio, or whatever—be causing low, long-term heating? I want to take a good look at the electrostatic or electromagnetic fields, and at some of the other fields around Saturn, as a way to get heat into the satellites—as such fields are different at different places, one or another of them may get around the question of there being less tidal heat at Enceladus than at Mimas."

Masursky had been spending much more time at the science-team meetings during this mission, he said, than the last, listening to the reports of the other teams on such matters as Saturn's magnetic field, its plasma activity, the radio waves it emitted, and so forth. There had been reports on radio waves emanating periodically from Saturn, apparently whenever a certain side of Saturn faced the Sun; on radio waves associated with some of the inner satellites; and on charged particles, possibly from Saturn's ionosphere, that were accelerated to high speeds in Saturn's magnetosphere, inside the orbit of Titan. One report that afternoon, by S. M. Krimigis of Johns Hopkins University, who was chief scientist of Voyager's low-energy-charged-particle experiment, had particularly caused Masursky to prick up his ears: it was of a cloud of fast-moving electrons encompassing the orbits of Rhea and Dione (527,828 kilometers to 379,074 kilometers from Saturn); heat, of course, involves the rapid motion of particles, and about a month afterward, Krimigis announced that, in strict physical terms—that is, with respect to the speed of the particles—the gas was the hottest known, ranging from six hundred million to one billion degrees Fahrenheit, or three hundred times hotter than the solar corona; the cloud was so sparse, consisting of only about thirty particles in a cubic foot, that anyone passing through it would hardly build up a bead of sweat—though what the cloud would do to a satellite orbiting permanently inside it might be another story. (It wouldn't have much effect, Masursky found out later—just enough, he told me, to indicate that the type of phenomena he was talking about was real.) And at the team meeting the next afternoon, Saturday, Masursky raised his hand high in the air when James Warwick, the head of the planetary-radio-astronomy team, who at the time of Voyager 1 had reported on the radio discharges that seemed to come from the rings, and who was the father of the putative Warwick Object, said he had heard the discharges again, though they were about ten times less frequent than before, the previous night. When Warwick called on Masursky, the geologist asked if the radio discharges might generate heat over long periods, warming satellites in the manner, possibly, of a microwave oven. "We've thought about radio fields doing that," Warwick said, "but the satellites are really much too far away to be affected."

As we got back to the imaging-team quarters, Masursky said, "Perhaps some of these fields were greater long ago than they are now. If so, the Earth and our Moon, as well as the moons of Saturn, might have been heated originally, in part at least, by some of these effects. I don't know what the answer is, but some physicists or geophysicists should take a look. About all I can say is, 'Here we have these small bodies, and they show unexplained signs of heat. And here we have these giant planets with a variety of extremely powerful fields. Could there be a connection?' "

Toward the end of the second encounter, the study of Saturn was entering a period of crosscurrents (such as the spacecraft itself seemed to have met as it rounded the planet): while some questions had been answered, and the data that might answer many others was in hand or almost so, a new group of questions—impossible to ask earlier—were just beginning to be formulated.

In the conference room next door to Bradford Smith's office, at around five in the afternoon, a number of imaging-team members, including Morrison, Johnson, Masursky, Soderblom, and Smith, were watching the television monitor, waiting for the resumption of pictures from Saturn. As had been the case for the last three days, one black, empty raw was still succeeding another. While they waited, the scientists were watching, on another television set, a closed-circuit NASA show about Saturn. They had tuned in midway through the film, which, using time-lapse photography and accompanied by sonorous organ music, depicted the construction of the deep-space antenna at Goldstone, California, where much of the Voyager data came in. As girder after girder was put in place, with the music rising in a crescendo, Smith said, "I hope they get it done by the time the picture comes in." The scientists laughed—it was the tension-easing laughter of people waiting on pins and needles for an event they are not entirely sure will happen.

Another blank raw appeared on the monitor.

The Goldstone dish, completed at last to a booming amen, was replaced by an interview with Garry Hunt, who was saying, "As Voyager passed by Jupiter, the planet's gravity sped it up, and at the same time the spacecraft's gravity slowed down Jupiter's rotation by one foot per trillion years."

"That's a good statistic," Torrence Johnson said. "I certainly hope it's right."

Another blank raw filled the monitor screen.

Tired of making small talk, the scientists were quiet. At last, Soderblom looked at his watch and said, "O.K., here we go."

The blank raw flicked off, and another apparently blank raw flicked on in its place. Soderblom stuck his head almost through the screen. "Hey, there's an image in there!" he said. "It's very faint. Dim the lights!" The faint curve of Saturn's illuminated crescent, with an even fainter curve sweeping outside it—the illuminated portion of the rings—sliced across the screen: a pair of abstract parabolas without meaning, unless you knew what they were.

The scientists yelled. They cheered. They pummeled each other on the back. They did all this despite the fact that they had just seen one of the worst pictures ever taken of Saturn, unenhanced, and visible only because the rings were reflecting light back onto the planet. ("Thank God for ringshine," said Smith.)

The spacecraft was now sixty-nine hours and about two million miles away from Saturn—almost back to the orbit of Iapetus.

One of the worst pictures ever taken of Saturn—and perhaps the most welcome: the first picture taken when the platform with the cameras was aimed back at the planet on August 28, 1981, when Voyager 2 was two million miles away, almost out to the orbit of Iapetus.

During the night, the platform was moved briefly, so that it could be exercised in what was called the Phoebe region—the area, in the 230° azimuth range, where that moon would be in six days' time, when Voyager would encounter it. Then, early that Saturday morning, the platform was brought back to 209°, pointing back to Saturn, and the narrow-angle, high-resolution camera took a picture. Apparently, all had gone so well that at Saturday's press conference Esker Davis was able to report that on Sunday morning they would most likely begin the high-resolution scan of the southern hemisphere of Saturn, more or less at the time for which it had originally been scheduled. Later, they would do yet another ring movie, also on schedule. "So we may be able to recover quite a bit of the postencounter science," he concluded.

After several months of analysis, the scan platform's problem would turn out to have nothing to do with insulation particles caught in gears. Using duplicate equipment, engineers had repeated every movement made since Voyager 2 left the Earth, and at almost the same point when the trouble had occurred in space, the platform on the ground failed. "It really seized," Laeser told me when I phoned him in November. The trouble was caused by the lubricant that was supposed to enable the gears to rotate freely around their shafts; it was ineffective. Those gear wheels that spun fastest, and had to revolve up to nine thousand times to effect a complete rotation of the platform, had become stuck to their shafts. The metal of the gears had abraded, and bits had gotten caught in the roughened surfaces of the shaft. The shaft itself, which was not meant to spin, had broken free at either end inside its housing and *was* spinning. "We are convinced that the same thing happened on the spacecraft, only not as severely," Laeser told me. "It's not a healthy situation to be in." It was no coincidence after all, he said, that the platform had broken down around the time of the ring-plane crossing, since that was the period of peak movement; when Verner Suomi suggested this possibility the morning after the breakdown, he had been quite right. The azimuth gearbox had been designed to last for four thousand rotations of the platform; instead, it had lasted through only three hundred and twenty—less than a tenth of its intended lifetime. Laeser is convinced that if the elevation gears had been used more, they would have broken down as well.

By Saturday morning the scientists were already reaping the rewards of having the cameras back on target. Now, with the Sun above the rings and Voyager below, the light and dark parts of the rings were reversed; they looked ghostly, as in a photographic negative. The Voyager 1 imagery had not been able to establish whether there were spokes on the

The underside of the rings, from 2.1 million miles away: Many of the light and dark features are reversed. For the first time, spokes are seen on the underside—white instead of black, because of forward-scattering.

An early picture of Saturn's southern hemisphere—though this one has been processed to bring out the rings, not features on the planet.

underside of the rings; now it could clearly be seen that there *were*—showing white against the dark B ring. This was the first discovery made after the cameras were fully turned on; the picture had come in about midnight the night before. It wasn't clear whether the spokes were simply the ones on the topside, seen through the rings from below, or were a separate, underside phenomenon. "I think we're seeing discrete spokes on the bottom side of the ring," Terrile said when I saw him. "The B ring is the thickest of the rings, and therefore the one we'd be least likely to see through." If that was the case (and it was later generally agreed that it was), then one theory of the spokes' origin—that they were caused by the charging (or, conversely, as another theory had it, by the *discharging*) of small dust particles by sunlight—was less persuasive, unless sunlight was penetrating the ring and doing the job. Correspondingly, the magnetic-field theory of their origin looked better, since the field existed on both sides of the rings.

A lot of new information would come from the postencounter pictures. Several days later, after I had returned to New York, an elated Ingersoll told me on the telephone that the high-resolution pictures of Saturn's southern hemisphere had turned out better than he had expected, and that the patterns in the south matched those in the north to such a precise degree that his coaxial-cylinder theory was more alive than ever. The alternate theory, that Saturn's winds were driven by heat from the Sun, was, of course, by no means ruled out.

Despite the procession of new pictures coming in, Jeffrey Cuzzi had spent Friday night going over the frames of the second spoke movie, looking within the rings for embedded moonlets. Next morning, he had nothing but reddened eyes to show for his search. I asked him if he felt

The southern hemisphere, imaged eight days after the encounter, and enhanced for surface features.

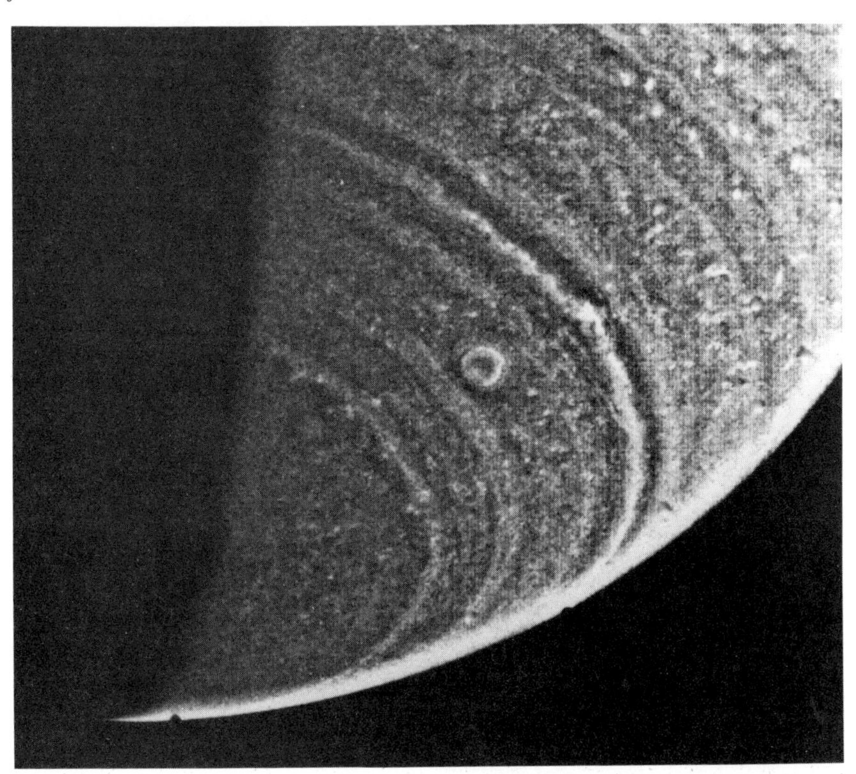

let down. "It's a disappointment not to have your eyes knocked out by a moonlet," he admitted. "But we'll do more processing—one may turn up yet. A lot of people believe in the moonlets."

The continued failure to find moonlets within the rings made the resonance theory look a little better. Indeed, at the moment, Richard Terrile was full of excitement. After looking at the dislocation of ringlets at the outer edge of the B ring, where it met the Cassini division, he had decided that it wasn't the so-called Mimas gap inside the Cassini division but the outer edge of the B ring itself that was the true Mimas resonance region. Moreover, he had evidence, in the case of one picture—there were four showing the dislocation—that Mimas had indeed been in the right place in relation to the distorted ringlets at the time the picture was taken. The distortion—a shift outward of a hundred and forty kilometers for the outermost ringlet—apparently halved as the distance inward was doubled, and this was consistent with a theory that Terrile's mentor, Peter Goldreich, had worked out previously. Not only had Terrile found the first visible demonstration that the resonance theory was valid but he had also very likely found the mechanism that

A picture of the outer edge of the A ring imaged on August 26, 1981, about half an hour before closest approach, when Voyager 2 was thirty-one thousand seven hundred miles from the rings; resolution is 2.5 miles. The linear features near the rim may be the product of resonances with some of the small new inner satellites.

held the outer edge of the B ring in place. Later, the photopolarimeter data showed that the outer edge of the B ring was sharply defined; according to Goldreich, this would be characteristic of a ring held in place by a resonance.

No embedded moonlets ever turned up in the main body of the rings. Bradford Smith told me a little later, "Our mistake regarding the moonlets may have been in having too strong a faith in a particular hypothesis; we were so certain we would see them—the only question was how big they would be—that we were caught with no alternative theory. We have to fall back on the resonance theory, which seems in some ways even less satisfactory." At the time, only two gaps had been definitely associated with resonances—the Mimas resonance holding back the B ring, forming thereby the inner wall of the inner gap in the Cassini division; and the Titan resonance, which coincided with a gap in the C ring that contained an elliptical ring. Later, though, others would be suggested. Allan Cook and Fred Franklin, a colleague of Cook's at the Smithsonian Astrophysical Observatory, whom I had met during the Voyager 1 encounter, suggested, tentatively, that there might be five resonances affecting the rings, plus seven more affecting the structure within the Cassini division. And Cuzzi and Pollack have suggested that some of the dozens of narrow gaps in the outermost part of the A ring might be the product of resonances with some of the small new inner satellites. Nonetheless, resonances could not account for more than ten percent of this sort of ring structure.

One reason for the strong faith in the embedded moonlets had been that they filled a need that now was far less urgent; not only was the photopolarimeter confirming what the radio-science team had suggested a few days before the encounter—that the A and B rings had virtually no gaps in them (other than a few known ones), at least down to a resolution of twenty kilometers—but it was continuing that observation down to the limit of its own resolution—a hundred and fifty meters—and finding that the A and B rings really *were*, for the most part, continuous discs of particles, with grooves and ridges rather than ringlets and gaps. As the presence of embedded moonlets had seemed the only means of creating absolutely clear gaps (Bradford Smith, for one, felt that resonances were incapable of this), the moonlets had now lost some of their purpose. Not all, however, for there were still a few clear gaps—two in the A ring, four in the Cassini division, and two more in the C ring—and the resonance theory could not account for them; it was still possible that they contained moonlets below the five- or ten-kilometer size. The photopolarimeter could well have missed these, because its

A rogue's gallery of some of Saturn's new small satellites, imaged by Voyager 2. The tiny ones at either end are the A-ring shepherd (far left) and Dione B (far right). Between them, the three pairs of satellites, arranged above and below each other, are, from left to right, the two F-ring shepherds, outer (above) and inner; the two co-orbitals, leading (above) and trailing; and the two moonlets near Tethys, trailing (above) and leading.

track through the rings was a single thin line that provided only a cross section. Differentiating at that size range between a moonlet and a big ring particle, however, would seem to be largely a matter of semantics—and such a small body wouldn't have been relevant to the theory anyway, because it would not have been big enough to clear the gaps. Though the ringlets had lost their independent status under the higher resolution, they had, as ridges, gained in number and complexity: what had been perhaps a thousand ringlets and gaps had become, in the high-resolution imagery, perhaps ten thousand features now known to be ridges and grooves, and many times that number under the far better photopolarimeter resolution. The bigger ridges—on the order of five hundred kilometers across, say—were, in the opinion of Terrile, probably permanent features, while those as small as ten kilometers across were very likely transient: ocean swell, perhaps, versus wavelets. Conceivably, they are part of the density-wave phenomenon, which many Voyager scientists increasingly see as a major force in the rings. Indeed, some months later, Cuzzi found a new type of transient wave within the rings, probably related to the density waves, which he has called bending waves. Like the density waves, they spiral, but unlike them they propagate inward instead of outward. More importantly, while the density waves are horizontal, the bending waves are vertical, as they are caused in part by the fact that Mimas and some of the other moons are somewhat out of plane with Saturn's ecliptic, moving above and below it; they literally bend the rings out of plane, causing gigantic warps seven hundred meters high, which move around the rings at the same pace as the satellite causing them.

Smith remarked that many of the ideas that had seemed good bets at the time of Voyager 1 apparently had to be revised later; that seemed to be the way with speculations—but he didn't think he or anyone else on the team was going to stop having them. A little later, Andrew Ingersoll said to me, "After this mission, I feel less embarrassed about speculation than I did before. In a sense, Voyager itself was just a great big speculation to start with—speculation that the spacecraft would last four years until it got to Saturn, speculation that we would find something interesting once we got there. Living through encounters like this makes you realize how much of science *is* speculation—just plain putting yourself at risk! Speculation precedes discovery. This business of going blindly forward—just going out there because you think you'll discover something, whatever it might be—I have a new respect for it."

On the day of my departure—Sunday, August 30th—when I dropped in to say good-bye to Terrile, I found him deeply engrossed in a picture of the rings on the interactive; it was a frame from the second spoke movie. "When I watched the movie, I noticed a gap in the Cassini division getting narrower and wider," he said. "It made me wonder whether there was a moon or a resonance in there pushing the gap in and out."

Clearly, the mission wasn't going to stop just because the spacecraft was leaving Saturn, any more than it had when the platform broke down. Indeed, whenever I talked with Terrile on the telephone in the coming months, the bulletins would pour in: Did I know that the photopolarimeter team had discovered in their data a tiny clear gap in the B ring after all, right where the Warwick Object was supposed to be? Did I know there was a third kinky ring in the big A-ring gap, lurking near one edge? Had I heard that several new moons had been discovered, some definitely and some only probably—one very likely near Mimas, found by members of the charged-particles team (it is no relation to the streaks near Mimas in the Voyager 1 pictures); one, and possibly two, just outside the orbit of Tethys, found in the imaging data by Stephen Synnott of JPL (it, or they, have no connection with the trailing Lagrange point, where Hansen and others had had their disappointment); one, definitely, near Dione B, also by Synnott; and another, definitely, between Dione and Rhea, by Synnott, too. ("Because I've been working with the imagery at the data-bit level for a long time, I've developed some new techniques and softwares others don't have," Synnott told me on the telephone when I asked him how he accounted for his success.) Together with Terrile's visual discovery of a moonlet within the F ring, Voyager 2 had discovered between four and six new satellites, all in the proximity of the major moons; these, together with all the previously known ones, made a grand total for Saturn of between twenty-one and twenty-three.

"It was a great encounter," Terrile said that day in August as I left. "I'm amazed that Saturn still holds so many secrets. There were many differences between Voyager One's view of Saturn and Voyager Two's. I'll bet if we go back again it will seem like a whole new planet. The differences show the value of going to a place like Saturn several times—I worry about going to Uranus with only one spacecraft. I'd love to see a Saturn orbiter mission, like the Galileo mission to Jupiter—one that might send a probe into Saturn's atmosphere, and maybe another into Titan's. It would be great to go back."

There was, of course, no hope in the foreseeable future for a third Saturn mission. In the coming weeks, the future of planetary exploration was thrown in jeopardy when the Reagan administration's Office of Management and Budget proposed, as part of the overall cuts in government spending, to curtail NASA's funds—representing less than one percent of the national budget—to such an extent that the money available to it would be sufficient only to cover operations of the space shuttle, which for reasons of national defense it was committed to carrying out. The Venus orbital mission to map that planet through its clouds, like the mission to intercept Halley's Comet, was scrapped altogether; the Galileo mission, on which the government had already spent three hundred million dollars, was in danger of being dropped as well. If the worst came to the worst—and in the two months after Voyager 2's Saturn encounter there was good reason to think that it would—funds for the further study of the Saturn data would be drastically cut, and the deep-space network of antennas around the world, which receive the data sent back by planetary spacecraft, would be shut down, so that when Voyager arrived at Uranus four years hence, the Earth would be paying no heed. In September and October, a number of the Voyager scientists—among them Smith, Soderblom, and Morrison—went to Washington to do what lobbying they could. One of their fears was the loss of future funding for the study of the Saturn data. Eventually, their rallying point became Galileo, which would be just enough to keep a minimum planetary program going, with a nucleus of technical talent. The lobbying was apparently successful, for funds were allocated in the 1983 budget for Galileo, and there would be funds for the deep-space network and for monitoring Voyager at Uranus and Neptune. There would also be funds for study of the Saturn data, though these funds were reduced. Later, funds were allocated for a much-curtailed Venus radar-mapping mission. The scientists, however, were not altogether easy. "You never can tell when some budget director, anxious to save money, will take the ax to these things again," Bradford Smith said to me on the telephone in December, when he was back in the office at Tucson.

Whatever budget problems occur on Earth, Voyager 2 will be going on to Uranus and Neptune; the spacecraft itself cannot be stopped by the Reagan administration or anyone else. (It won't be stopped by the platform problem, either; Richard Laeser told me over the phone in November that although the scientists and engineers might not be able to move the platform as freely at Uranus as they had at Saturn and Jupiter, all the mission objectives could probably be met by orienting the spacecraft itself to aim the platform instruments and by writing the sequence for targeting in such a way as to reduce the number of movements.) When I stopped in to say good-bye to Bradford Smith at the end of August, he was full of plans for holding the imaging team together: perhaps, during the next four years, he could scrape up the funding to assemble everyone at least once a year; every spring, Uranus comes within view of telescopes on Earth, and that, he thought, would be a good time to meet and go over the latest targeting information. Every member of the team had told him that he or she wanted to be back at JPL for the Uranus encounter in 1986. Indeed, that planet was already well to the fore in everyone's minds now—there were Uranus jokes and Uranus cartoons on the bulletin board; one cartoon showed Voyager, its broken platform in a sling, carrying a suitcase labeled "Uranus or Bust."

Far less is known about Uranus than was known about Saturn before Voyager visited it. Uranus and Neptune, though they are generally considered giant planets, are much smaller than Saturn and Jupiter; each is a little over thirty thousand miles in diameter, and has only five or six percent of Jupiter's mass. I asked Smith what he was most looking forward to when Voyager got to Uranus. Much of the planet's attraction, he told me, lay in its remoteness, its mysteriousness, and its contrariness. "We don't know whether it rotates in sixteen hours or twenty-four," he said. "It has no internal heat source—though Neptune, which is about the same size, does. Uranus is a complete anomaly. And since it rolls around its orbit on its side, as though its rings were wheels, it must have the oddest weather in the solar system." The nine Uranian rings, which were not discovered until 1977, are very narrow; one, the Epsilon ring, is only forty or fifty miles across. They're also very dark; not only do they most likely contain carbonaceous material but, since the temperature is much colder at Uranus than at Saturn, a greater quantity of all sorts of other volatile compounds must have condensed as well. Evidently, there is a considerable amount of methane in the atmosphere of the planet itself—turning it a shimmering opalescent green—and who knows what other exotic materials may lie trapped in the ice of Uranus's five known moons—Miranda, Ariel, Umbriel, Titania, and Oberon? As for Neptune, even less is known about it, its rings (conjectured from earlier data in the spring of 1982), or its two known satellites, Triton and Nereid. Triton is thought to be a rocky body with a thin atmosphere containing methane; Nereid is probably icy. Neptune itself is covered much of the time with a haze. The planet is a delicate light blue, because of the methane—though there are some nonscientists who believe it is that color because it is just plain cold.

As Voyager 2 moves on to worlds unknown to the ancients and to many moderns as well—Uranus was found only in 1781 and Neptune as recently as 1848—the scientists will go on analyzing the Saturn data, refining their theories, and very likely making new discoveries; the process has only just begun. Already, though, with Voyager they have learned far more about Saturn, its rings, and its moons than had been known in all previous history. If to know a thing is to claim it in some way, then this odd, icy outpost—like the planets nearer the Sun—is now more nearly ours.